CYCLES OF NATURE

An Introduction to Biological Rhythms

Andrew Ahlgren
Franz Halberg

National Science Teachers Association

National Science Teachers Association
1742 Connecticut Avenue, N.W.
Washington, D.C. 20009

Library of Congress Catalog Card Number 89–063723
ISBN Number 0–87355–089–7
Printed in the United States of America
First edition

CYCLES OF NATURE

CONTENTS

About the Authors

Andrew Ahlgren is professor of education at the University of Minnesota and associate director of the university's Center for Educational Development. Currently he is on leave at the American Association for the Advancement of Science as associate project director in the Office of Science and Technology Education. He has worked with Franz Halberg off and on for twenty years to clarify and disseminate the ideas of chronobiology. The line of argument he has drawn in this book arises largely from Halberg's pacemaking perspective on theory and practice but, as an elementary introduction, emphasizes only some of the central ideas from a rich and rapidly growing field.

Franz Halberg is professor of laboratory medicine and pathology at the University of Minnesota and is director of the university's Chronobiology Laboratories. Although he has contributed extensively for forty years to almost all aspects of rhythm study (indeed, he coined the terms "circadian" and "chronobiology"), his special focus is on how the understanding of rhythms can be applied to improving human health and well being. To this end, he has helped to shape how chronobiological research on biological rhythms is done, interpreted, and used and to establish it in biomedical research centers around the world.

Foreword

In the traditional logic of biology, if a process is always observed and measured under identical circumstances, such as temperature and chemical environment, the process will occur at the same rate each time. And if measurements are made very close together, or at about the same time each day, this logic may prove correct.

But modern studies have shown that, more generally, biological measurements are not usually the same. To everyone's surprise, studies made under identical conditions but at different times of day revealed that another variable was involved: Rates varied as a function of the time of day! That is, they varied cyclically, with a period of about 24 hours. By way of example, all other things being equal, the rate at which the human liver detoxifies 50 ml of alcohol consumed at sunrise is quite different from its speed if the same amount of alcohol were to be consumed at sunset. (The human liver is much more effective if the alcohol is administered in the early evening.) But the liver's response rate is about the same every morning, just as all evening response rates are similar to one another. Thus the liver acts as if it can tell time.

Furthermore—and this is what makes the field of **chronobiology** (which this approach to biology has come to be called) so exciting—as unexpected as it might seem, this periodicity persists even when an investigator eliminates all of the obvious clues to time of day. The cycles are innate.

These persistent daily cycles, expressed by almost all plants and animals, are found even when human beings are the subjects for the experiments. The unavoidable conclusion is that within all life there are *pacemaker* systems. Some of these, found in eukaryotes (organisms whose cells contain a distinct nucleus), routinely measure off intervals of about a day (a month, or a year), and these systems mold most fundamental biological activity into predictable time intervals. That story is the theme of this book.

Given that these timers operate almost everywhere, in almost all living things, scientists must seek to discover more about them. Thus far, while timing systems have been found in an expanding list of animals and plants, far too little is known about how they work. The field of chronobiology has arisen to tackle this problem. Its discoveries to date fascinate everyone who learns of them; thus the subject matter is ideal for exploration in the classroom. Unfortunately, the lion's share of the published work on the subject can be found only in scientific journals. Another factor limiting public access is the fact that through the years, chronobiologists have adopted a highly technical vocabulary to describe their findings precisely. This habit, common in most new fields, becomes an obstacle to easy understanding and greatly reduces the dissemination of the field's subject matter, keeping it not only from the public and the classroom, but from other scientists as well.

The book you now hold in your hands, however, is a well thought out prescription designed to remedy this condition. The writing is clear: Some of the best, most important, and most interesting material has been drawn from a vast literature and presented here for beginners. And since seeing is believing, the last chapter is a series of *hands-on* laboratory exercises. Students who actually carry out these observations will be well rewarded—something that I have discovered from my own work on rhythms. Although I have studied a great variety of plants and animals over the last 25 years, each time I study a new organism under controlled, *timeless* conditions and watch it wake with a rising sun that it cannot see or feel, I am enthralled and filled with the wonder of nature.

John D. Palmer is a researcher in the area of chronobiology and author of a college text on chronobiology: *An Introduction to Biological Rhythms.* New York: Academic Press, Inc. (1976).

John D. Palmer

Department of Zoology
University of Massachusetts
Amherst, Massachusetts

Acknowledgments

A project like *Cycles of Nature: An Introduction to Biological Rhythms* does not reach publication without the contributions and support of many people.

This manuscript was reviewed by John F. Carroll and Dora K. Hayes of Beltsville Agricultural Research Center, Beltsville, Maryland; Rita Hoots of Woodland, California; Lawrence Gilbert of Brooklandville, Maryland; and Barbara Schulz of Seattle, Washington. They not only verified scientific accuracy, but evaluated the presentation of the concepts and the ease and safety with which the activities can be performed. Their efforts are greatly appreciated.

Special thanks go to Dr. Erhard Haus, chief of pathology, St. Paul Ramsey Medical Center, who reviewed early drafts; Julie Dobbs, St. Paul Public Schools, who piloted early drafts with her students; and Glenn Radde, Minnesota Department of Natural Resources, who analyzed the results of further pilot tests.

Cycles of Nature was produced by NSTA Special Publications, Shirley Watt Ireton, managing editor; Cheryle L. Shaffer, assistant editor; Ward Merritt, assistant editor; Michael Shackleford, assistant editor; Phyllis Marcuccio, director of publications. Cheryle Shaffer was NSTA editor for *Cycles of Nature*. At AURAS Design, the book was designed by Rob Sugar, and production was handled by Sylvie Abecassis. Charts were created by Mark Associates. The contributions of all these people were essential in the preparation of this book. Early versions of the manuscript were written under a grant from the Educational Directorate of the National Science Foundation to the Chronobiology Laboratories, University of Minnesota.

INTRODUCTION

Around and Around: The Importance of Cycles

This book is an outline for a short (one- or two-week) study of chronobiology, a new field of science that explores the relationships between time and biological functions. It develops step-by-step the reasoning that leads to the current scientific understanding of biological rhythms. The unit can easily be inserted into a standard middle or high school biology course.

Because the scientific study of biological rhythms begins with data, Chapter One provides a brief review of the ways to collect, graph, and interpret data. Chapter Two introduces some of the cycles in nature, especially those of the human body—from dream cycles to menstruation to body temperature. Many cycles will be introduced, mapped out, and explained. Chapter Three explores how these cycles come about and explains the differences between external (environmental) and internal (purely biological) influences. Chapter Four explores the internal workings of organisms to determine whether there is a single master source of timing information that synchronizes an organism's many interacting cycles. Chapter Five discusses the impact of rhythms on society and asks how an understanding of them could bring progress in medicine, work schedules, and everyday life. Chapter Six offers a brief historical perspective on the study of biological rhythms, and Chapter Seven outlines activities to demonstrate cycles in chemicals, plants, and animals.

But why learn about cycles? Are they really an important

part of studying biology? Yes—the relationship between time and biological functions seems important to all life. Cycles occur in all living things, as well as isolated tissues and chemicals.

A good way to begin may be with a surprising demonstration. It involves six readily available chemicals: hydrogen peroxide, potassium iodate, sulfuric acid, malonic acid, soluble starch, and manganese sulfate. When these chemicals are combined in the correct amounts, the solution turns from amber to blue-black to colorless in cycles of about 15 seconds, for several minutes. (Directions for performing this demonstration are given in Chapter Seven.)

Many other cycles in nature are equally visible, although their inner causes are hidden from us. Humans operate on cycles as well. In one experiment on kidney cycles, men living in the arctic, under conditions of nearly constant light and temperature, were given special watches which, unknown to them, ticked off a 21-hour day. After several weeks of living under these conditions, the explorers, upon returning to civilization, were surprised to learn that the "days" to which they had been accustomed were not 24-hour days. Analysis of the data, including the chemical analysis of frequent urine samples, showed that different biological cycles responded in different ways. The usual 24-hour cycle of urine acidity quickly shortened to 21 hours. The cycle for potassium, however, stayed close to 24 hours. And sodium showed a combination of 21-hour and 24-hour cycles.

Another dramatic exhibition of cycles involved four ground squirrels in a laboratory. These squirrels, raised in the lab from birth, in constant darkness and with no contact with the outside world, *still hibernated each winter and awoke each spring*. Each year of the four-year study, the squirrels' hibernation schedule varied by only a few weeks from those of squirrels in the wild, showing that they were following their own built-in schedules.

In order to expand the reader's knowledge of cycles beyond the regular text, there are two kinds of questions throughout this book. Those in the margins are meant to provoke individual thought and class discussion; they have no right or wrong answers. (Margins are also used for chapter introductions and additional information.) Those questions at the ends of the chapters are geared specifically toward the material and can be assigned as homework.

CHAPTER ONE

Seeing Cycles

A cycle, plainly put, is a pattern that repeats itself continually. But when can we look at a set of data and confidently assert that "this shows a cycle?" We will find it has something to do with how confident we are that the small amount of information we *have* is truly representative of the total information that there *is*.

This section addresses two questions: What is a cycle? and What is convincing evidence for the existence of a cycle?

Measuring the Tide

If you have ever been to the ocean for any length of time, you have probably noticed the tide—the gradual rising and falling of the water's surface due to the gravitational pull of the moon and sun. Suppose you had to figure out when the tide comes in and when it goes out, with no help from outside sources. What would you do?

Consider Jessica, a young girl who lived by the beach, who decided to do just that. Jessica liked to dig for clams, but she could do it only when the tide was out. One morning, Jessica found a long stick on the beach and marked numbers on it to measure the height of the water. (She calibrated the stick from 0 to 100 cm because she knew from experience that the tide never got any higher than about 1 meter.) Then she drove the stick into the sandy bottom where she remembered having seen the water's edge the farthest out. The water came up to about the 20 cm mark (Figure 1.1, Day 1). She came back about noon and found the water up to 38 cm, less than a foot higher. She didn't look again until the next morning, again about 8 o'clock, and found the water

In this example the pattern of the tide and the calmness of the water have been greatly idealized. Tides have a complex pattern that results from cyclic changes in the position of the moon, earth, and sun and from the shape of the local ocean bottom.

at about 36 cm, almost the same as the previous noon (Figure **1.1**, Day 2). When she checked again at noon, the water had dropped to just 22 cm. Now she was puzzled—the water had risen by almost 10 cm during the first morning, but dropped by almost 10 cm the second morning.

Jessica decided she would have to watch the tide much more carefully to figure out what was going on. So the next day, Jessica got up at 7 o'clock, had breakfast, packed a lunch, got a pencil, her notebook and her wristwatch, and went to the beach. She sat there until evening, in the same spot, reading a book, throwing rocks into the ocean, watching sea gulls, and each hour recording

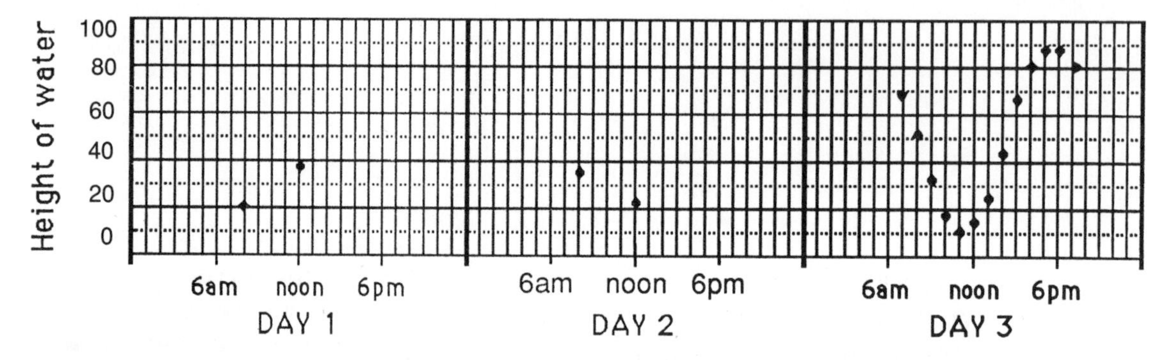

Figure 1.1 Jessica's measurements of the tide over three days. On days one and two, she measured only at 8 a.m. and noon, but on day three, she took hourly measurements.

the water level as measured by the stick. At the end of the day she returned home tired, a little sunburned, and with sand in her shoes. At home, she uncrumpled the sheet of paper with the numbers and times carefully recorded on it and made a new graph (Figure **1.1**, Day 3).

Jessica now saw the unmistakable cycle of the tide. The water would fall gradually until it reached the lowest point, and then it would begin to rise. About six hours later, it would reach its highest point and then begin to fall again. But why had she gotten those reversed results on the first 2 mornings? She decided to continue her watch on the beach for a few days (Figure **1.2**, Days 4, 5, and 6).

Figure 1.2 Jessica's hourly measurements of the tides for days 4, 5, and 6

Jessica could imagine a regular swing in and out with almost

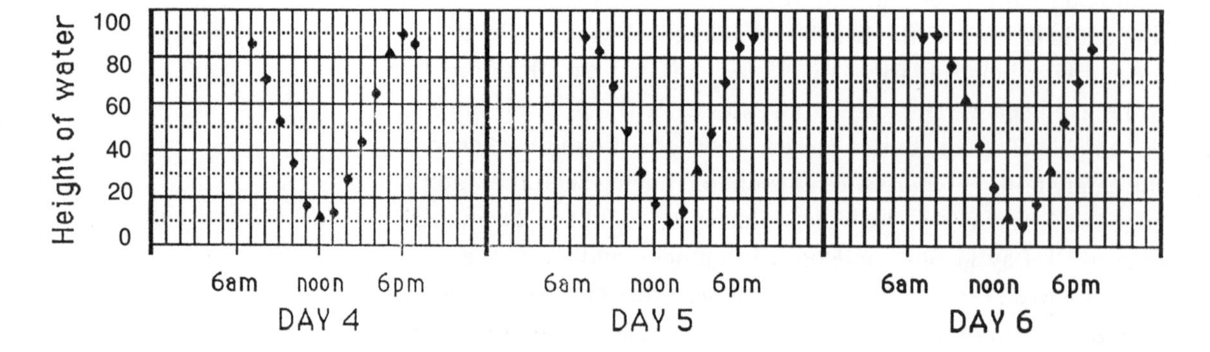

13 hours between successive highs or successive lows. Presumably there was another high or low at night, but she hadn't felt like going down to the beach then. Now she wished that she had checked just once. (maybe next week.) With that idea she could make a sketch for all six days like Figure **1.3**, showing how the slightly changing time of low tide had caused the reversal in her measurements on the first two days. Jessica did not know what caused the tides; she just knew that it came in and out slightly later each day. She wasn't sure how far ahead she could predict, but over a few days it seemed that she wouldn't be off by more than half an hour. And that was what she needed to know to dig for clams!

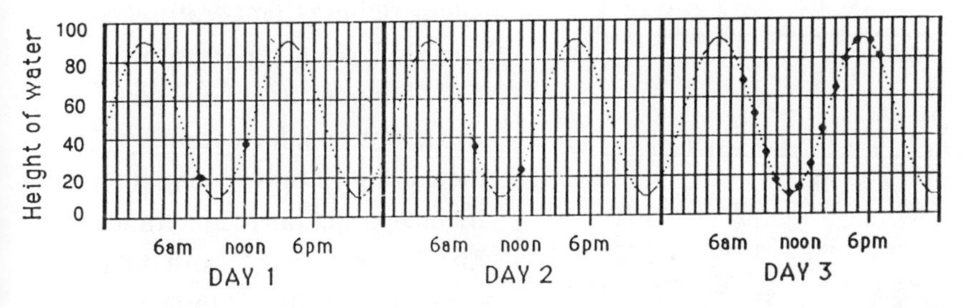

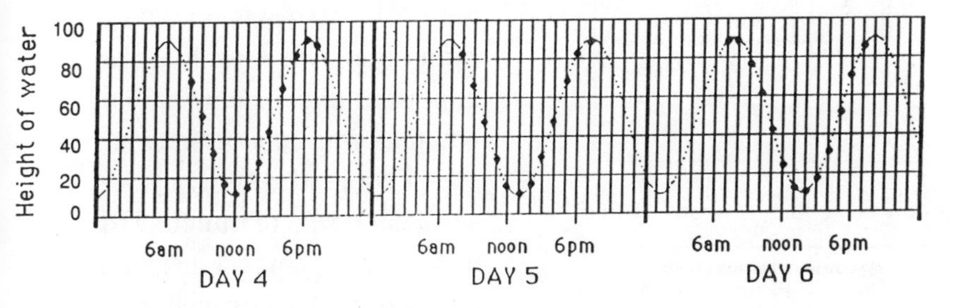

Figure 1.3 Jessica's graph of all six days of measurements, with a curve representing estimated tide heights throughout each day

So that's what Jessica learned about tides. But what did she learn about investigating cycles? Why, for instance, did she stop measuring when she did? Why didn't she stop after making the first graph, or for that matter, take still more readings after the days on the beach? Was one measurement per hour often enough to check the timing of the tide?

By looking at all of the graphs together, we can see that although Jessica's first measurements were accurate, they were not fully representative: They did not tell her enough to predict when the next high tide would come. Those first measurements were like reading every tenth page of a book instead of every page. The tenth pages alone, although part of the book, are not representative of the whole book.

And why not measure even more often than once every hour? Jessica believed that there was a steady motion of the tide from one hour to the next and nothing special happened in between. She did not have to measure more often than once an hour, at least if predicting low tide to the nearest hour would be close enough. But why was she not satisfied with her first measurements? What were they lacking?

Whether you are studying cycles in the tide or anything else, it is important to have enough data. If you don't, a true cycle can look like random changes—or random changes may look like a cycle. As Jessica learned, you can find out more by measuring more often. For example, consider the daily rising and falling of body temperature—if you are looking for your own daily temperature cycle, you will get more information by measuring 12 times a day than by measuring only three times. Another way to get more data is to measure over a longer time, for instance two weeks rather than two days. It is also important to consider *when* you take your measurements.

Figure 1.4 The characteristics that describe a sine curve

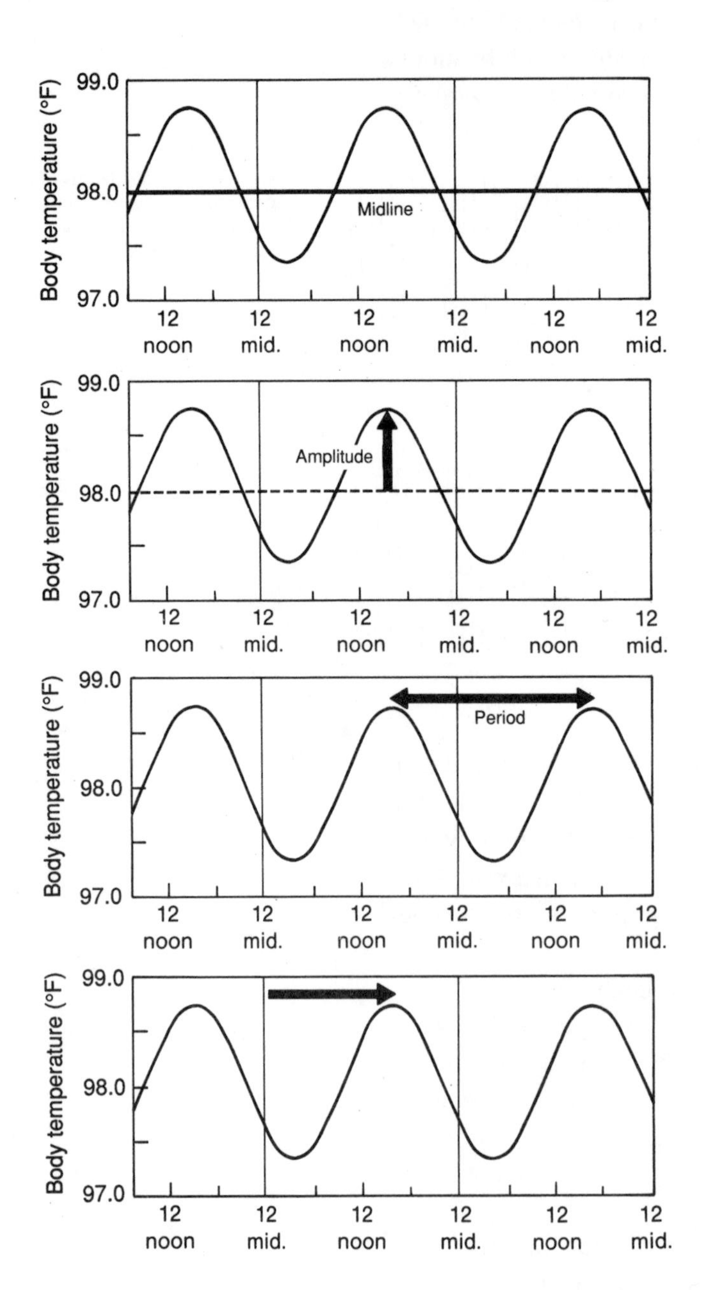

People who study cycles say a good rule of thumb is to make at least several measurements per cycle at roughly equal intervals, over several cycles. To look for hourly cycles, for example, measure at least six times an hour for several hours. To look for daily cycles, measure at least six times a day (about 4 hours apart) for several days. To look for about-monthly cycles, make at least six measurements a month for several months, and so on.

Some cycles need more data than others before you can spot them. Body temperature, for example, shows a fairly smooth, regular variation, so a few measurements a day will suffice. On the other hand, heart rate varies less smoothly, depending on what your activities and thoughts are. The smooth change of a daily cycle may be half hidden by these short-term changes. Like heart rate, blood pressure also shows many short-term variations, so it requires more measurements to show this cycle clearly. Perhaps a dozen measurements a day over several days would be necessary before the details of the gradual rising and falling become apparent.

There is another vital aspect of measurement: being careful. Take the time to make the measurement accurately. Arrange for the least possible disturbance before and during measurements to give short-term influences a chance to settle down. For best results in investigating biological cycles, keep a regular sleeping and eating schedule during the measurement span and also for a number of cycles before even beginning measurements.

A Short Math Lesson

The general shape of the curve that Jessica got when graphing her tide cycle often fits biological cycles fairly well: It changes direction gradually around the peak and trough and changes more rapidly in between. A simple mathematical function that behaves like this is called a sine curve. A sine curve is fitted through Jessica's data in Figure **1.3**. Sine curves are important in the study of cycles and, although we will not be studying them in depth, it is important to know that they are mathematically very simple. Any sine curve can be described on a graph by only four numbers, as shown by the sine curve graphing body temperature over time in Figure **1.4**: the **midline**, a middle value through which the curve rises and sinks; the **amplitude**, the difference between the midline and the highest (or lowest) point on the curve; the **period**, the interval between any like parts of the curve (peak to peak, trough to trough, etc.); and the **peak time**, when the sine curve reaches its highest value.

As you study cycles, you will gather and plot data on a graph and see what sine curve reasonably fits the data. You shouldn't expect that all the data will exactly match any simple curve. Some fine points to note: The midline is not the same as the average of the data unless the data are taken at equal intervals around the clock; the full range of cyclic variation from peak to trough is twice the amplitude; the highest actual measurement may not occur at the same time as the peak of the best-fit sine curve. For example, Figure **1.5** shows measurements of body temperature that were taken by a student over three days for a school project. Figure **1.6** shows, in addition, the sine curve that best fits the whole set of temperature data, although it does not actually go through any one of the data points. The **midline** value is 98.6°F, the **amplitude** is 0.8°F (above and below the midline), and the cycle peak occurs 16 hours after midnight, about 4 p.m. There will be momentary variations, there will sometimes be errors in measurement, and the underlying cycle may have a more subtle variation in time than a simple sine curve. However, the simple sine curve often turns out to be a convenient summary of a cycle. The sine curve description of a cycle can be readily compared to descriptions of other cycles: those for other variables, other individuals, or other kinds of organisms.

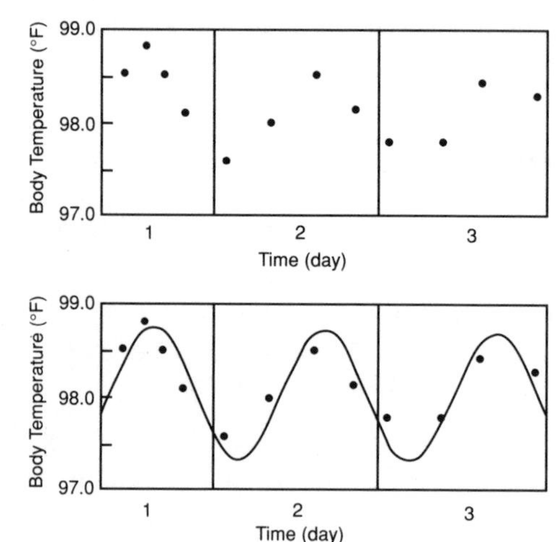

Figure 1.5 Body temperature measurements made over three days

Figure 1.6 Best-fit sine curve (with 24-hour period) for data in Figure 1.5

Study Questions

1. What is a cycle?

2. Jessica's initial measurements did not indicate a cycle. Her later measurements did. Explain.

3. When looking for a cycle, how many measurements should you take during each cycle, and over how long a span of time?

4. List the four characteristics that describe a **sine** curve, and briefly explain each one. Label these four characteristics for the curves in Figure **1.3** and Figure **1.6**.

5. In doing an experiment, if the first data you got indicated a cycle, how certain would you be that there was one there? Explain.

CHAPTER TWO

Putting Cycles on the Map

Consider dreaming. Most dreaming occurs while the sleeper's eyes are shifting rapidly back and forth under the eyelids. This is called "rapid eye movement" sleep, or REM sleep. Figure **2.1** is a graph of the percentage of time one person spent in REM sleep, averaged over 30 nights. The horizontal scale is not the actual time of night, but the number of hours after the time the subject went to sleep (which varied from night to night). You can see that the graph swings up and down, showing that the amount of dreaming was uneven. During the first half of sleep, dreaming occurred in dis-

We will now look at other situations which, like the tide, reveal cyclical patterns. Although the tide may be well explained, other cycles may not be; for although the cycles themselves may be visible, their underlying causes often remain hidden from us, and it is only through careful, thorough research that we can unlock their secrets.

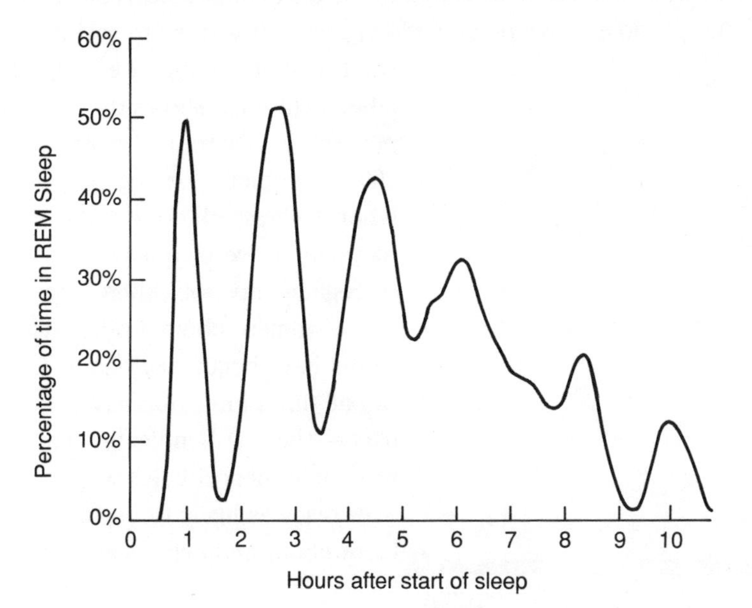

Figure 2.1 Percentage of time spent in REM sleep (smooth curve drawn through data points averaged over 30 nights)

tinct cycles, about two hours apart.

In this chapter, we will introduce some of the many distinct cycles in the human body. Some of these cycles, such as the menstrual cycle, are commonly known; however, the majority of cycles are unrecognized by most people.

Let's look at another common cycle. Even though you may have been taught that *normal* body temperature is 37.0°C (98.6°F), your temperature actually swings up and down by as much as a degree or more every day. Such change is evident in Figure **2.2**, a graph of a man's temperature measured automatically every

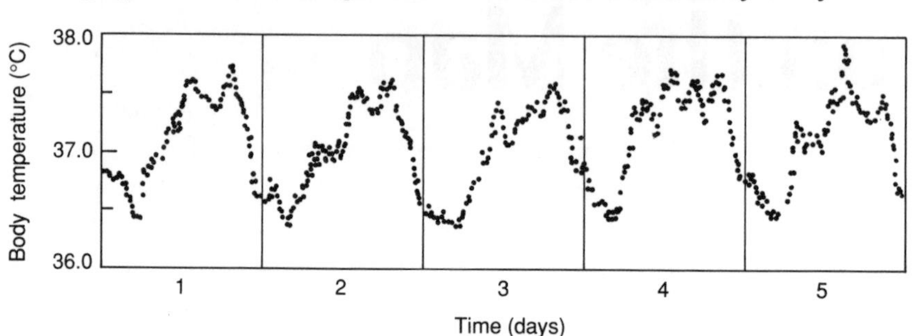

Figure 2.2 Man's body temperature recorded automatically over five days

15 minutes over five days. Although the temperature pattern wasn't exactly the same every day, you can see that the man's temperature did not remain at 37.0°C. His temperature measurements show a distinct cyclical pattern.

Found in the Crowd

Now consider, instead of an individual, a group of people. Historically, many studies have been made of pregnant women to see at what time of day they went into labor. Figure **2.3** shows the percentage of a total of over 200,000 women who began labor during each hour of the day. If labor were equally likely to begin at any hour, then 1/24 of the women (about 4 percent) would have begun each hour. But all times aren't equally likely—the graph indicates that women in these studies were over twice as likely to begin labor about midnight as at

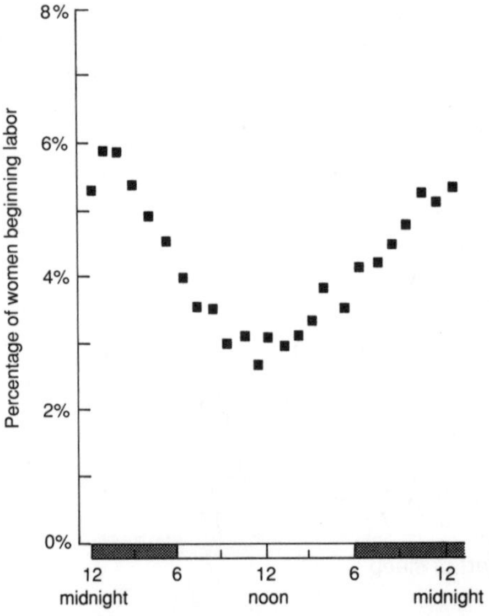

Figure 2.3 Percentage of 200,000 women beginning labor during each hour of the day

noon. In between these times, the likelihood rose and fell along a smooth line.

Because most women bear no more than a few children, it is unlikely that the labor-onset cycle of the human female population will be illustrated by the experience of any one woman. We must examine a large number of women in order to see the pattern. For the same reason, we can't talk about an individual insect's hatching cycle or an individual cell's division cycle. We observe these cycles only when we collect data on a fairly large population.

The existence of a population cycle reveals an underlying tendency in members of that population. Individual insect eggs have cycles in their metabolism that cause the probability of hatching to vary at different times of day. Individual women have underlying hormone cycles that, when gestation is complete, made labor easy to trigger during the early hours of sleep. When we observe a population cycle in once-in-a-lifetime individual events, we believe that it results from an underlying rhythm of sensitivity in individuals.

A Day in the Life

Figure **2.4** shows the peak times for 21 of the many cycles that have been demonstrated in humans. The data are taken from

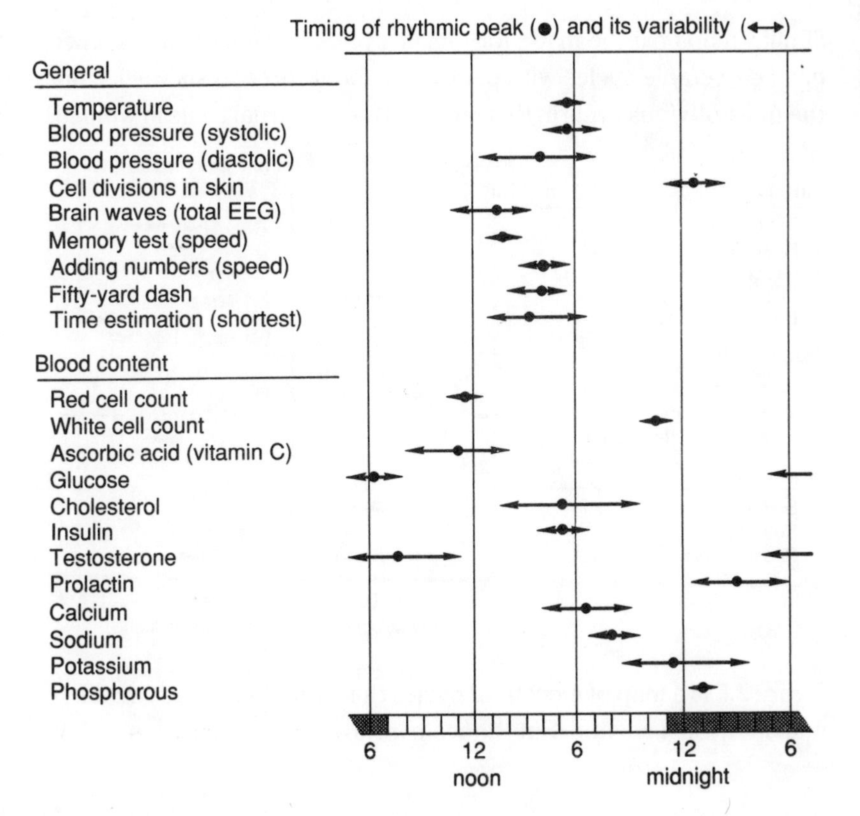

Figure 2.4 Common peak times for 21 of the many circadian cycles in the human body

Human hormone concentrations rise and fall in periods of several weeks. One study of this phenomenon was done by a man in France who spent four months living in a cave, with no light or temperature clues from the outside environment. Every morning after shaving, he collected the stubble and put it into an envelope. He then weighed the envelopes and found that the amount of stubble—and hence the output of the hormone which causes beard growth—followed an about-three-week cycle.

many different studies, with the scale across the bottom giving the hour of the day, and the darkened section of the scale showing the approximate hours of sleep for the people in the studies. The black dots indicate the average peak time of the cycle for each group, and the arrows show the usual range of variation among people.

The cycle peaks for most of the variables in the "general" list occur between noon and 6 p.m.—in the middle of the waking hours. These variables correspond to task performance and coordination. Temperature and blood pressure are related to the rate at which your body functions can work, while the other four are indicators of actual performance. (The exception among the "general" variables is the rate of cell division in skin, which peaks soon after the beginning of sleep. Cell division does not directly contribute to performance, but is part of growth and repair.) Variables in the lower list are for some of the contents of blood and have peak times for cycles across all 24 hours of the day. You can see that the cycles in red and white blood cell counts are opposite—the peak times are 12 hours apart.

All of the cycles in Figure **2.4** are called "circadian"—Latin for "about daily"—cycles. (Later you will see why the "about" is important.)

Longer than a Day

Similar maps can be made for longer cycles. Both men and women have detectable cycles with periods of about three to six weeks, but the most obvious cycle in this range is the menstrual cycle in women.

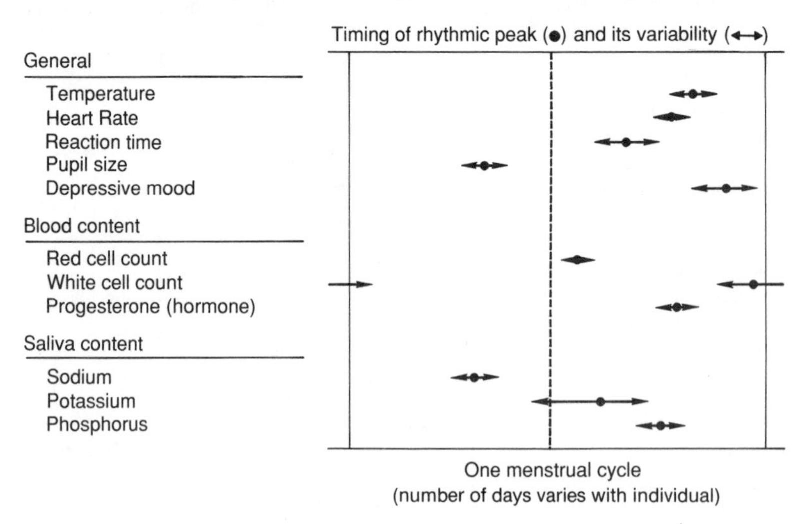

Figure 2.5 Common peak times for 11 about-monthly cycles (relative to the length of each woman's menstrual cycle)

Figure **2.5** is a map of menstrual cycles that is similar to the circadian map in the last section. But the scale is now several weeks instead of

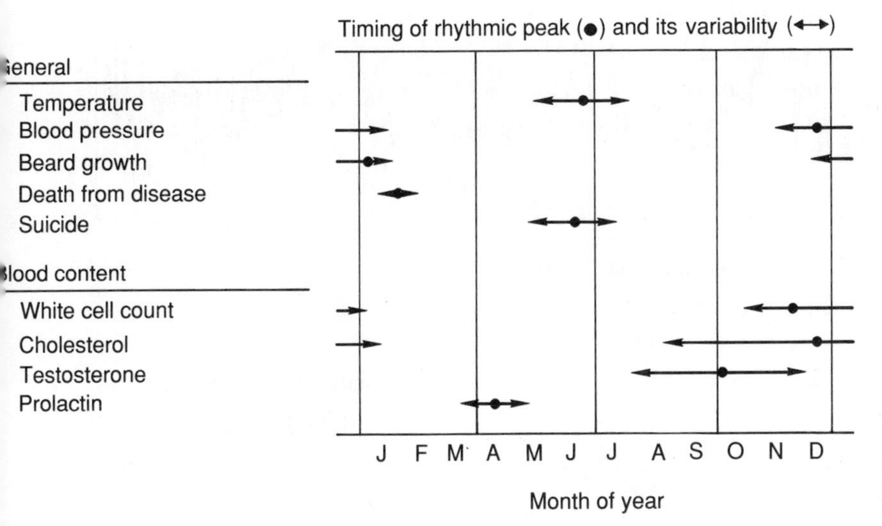

Figure 2.6 Common peak times for nine about-yearly cycles in the human body

just a single day. (Because women have different cycle lengths, there is no particular number of days given—the time axis is one complete cycle.) The dots show, on average, when in the cycle each peak occurs. For example, whatever the length of the cycle, whether 26 or 31 days or some number in between, the peak for red cell count occurs right in the middle of the cycle.

Figure **2.6** is similar to the previous maps, but it is based on only a few individuals. The time scale is now a year. The dots show the cycle peaks of each variable along a time axis from January through December. The hormone prolactin, for example, peaks in early spring, while the hormone testosterone reaches its peak in early fall.

All three of these maps illustrate the point that biological variables are cyclical, and that different variables can peak at different times.

Since there are also variations in cycle peaks among individuals, it is not always wise to make estimates based on information taken from measurements of groups of people. However, there are some kinds of cycles, such as the labor-onset cycle, that can be seen *only* in a collection of people.

Different cycles for the same biological variable can occur simultaneously in an organism. For example, the circadian temperature cycle in human beings and the temperature cycle which accompanies the female menstrual cycle can actually combine to create a pattern of change that includes both cycles. Figure **2.7** is a record of a woman's body temperature taken three times a day for four months. You can see both the circadian variation and the

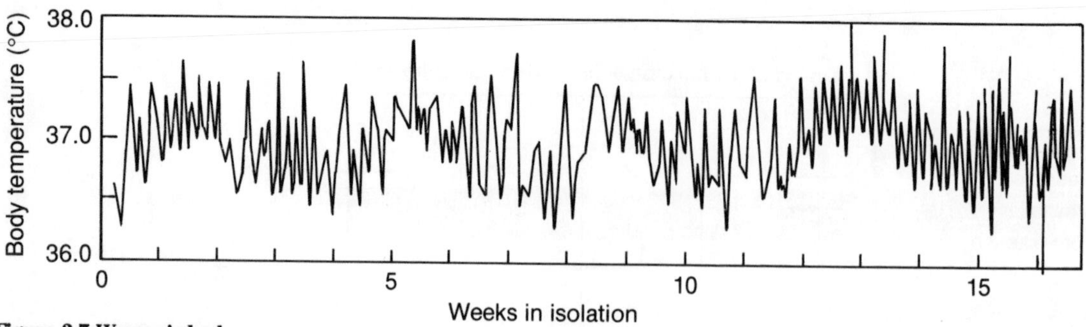

Figure 2.7 Woman's body temperature rhythm measured several times daily over four months of isolation in a cave

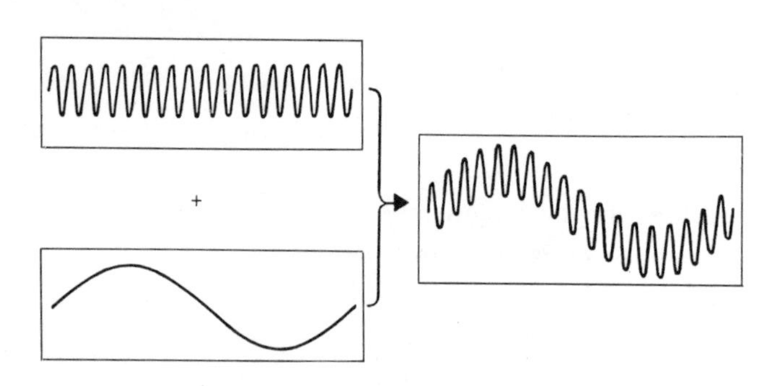

Figure 2.8 Theoretical composite of a circadian temperature cycle and the temperature cycle which accompanies the menstrual cycle: two cycles with different periods

swing up and down of its midline in a cycle of about four weeks. This example of interplay among cycles resembles the algebraic sum of their variations in the separate cycles (see Figure **2.8**).

Biological processes are influenced by many cycles, from those only fractions of a second in period to those with periods of about a week, several weeks, a year, and even longer. The next chapter explores the interactions between cycles and the forces that influence their functioning.

Study Questions

1. Dreams seem to appear in a cycle during a person's rest span. What is the period of this cycle? How many peaks occur within a typical night's sleep?

2. At what time of day was a pregnant woman found to be most likely to begin labor? At what time was she least likely to begin?

3. Define the word "circadian" and explain where it comes from.

4. Give five examples of bodily cycles. Which are circadian cycles? Which are longer or shorter than circadian cycles?

5. What is meant by the phrase "population cycle?" How does it relate to an individual cycle?

6. Population cycles affect death as well as birth. The occurrence of strokes in the U.S. shows an about 3 1/2-day cycle and an about 7-day cycle as well as a circadian cycle and an about-yearly cycle. Draw the pattern that would result from the combination of these three cycles.

CHAPTER THREE

What Drives Cycles?

The Environment's Role

In the preceding chapter we were introduced to some of the many cycles in nature, and we learned that different rhythms within an organism peak at different times. In this chapter we will look more closely at these cycles and try to determine what factors are responsible for giving rhythms their distinguishing characteristics.

Because the period of a cycle in an organism so often matches the period of an environmental cycle, the most obvious answer to the question "What drives biological cycles?" is that they are driven by cycles in the environment. According to this idea, we would imagine that flowers open every day because the sun shines on them, animals hibernate in late fall every year because the weather gets cold, and birds migrate every spring and fall because day length changes. Do they? Let's look at several lines of evidence.

Schedule Shifts

The influence on cycles by the environment can be demonstrated clearly by shifting the lighting or temperature schedule of an organism's environment. For example, hamsters become active at dark every evening (possibly running on a wheel) and retire to their nests when it gets light every morning. Imagine that a hamster is kept on a normal schedule of day and night—a light over its cage is turned on from 6 a.m. to 6 p.m. and is left off from 6 p.m. to 6 a.m. If kept on this schedule for a week, the animal will follow a standard pattern for hamster life, starting to run at about 6 p.m. and going back to sleep at 6 a.m. If the lighting schedule is then changed so that the lights go on and off three hours later (light

There are, of course, steady trends in biological variables as well as cycles. In humans, height, elasticity of tissue, and stiffness of the lens of the eye are variables which generally progress in one direction. How many others can you think of?

from 9 a.m. to 9 p.m.), the hamster will soon shift its own schedule to match.

Many leafy plants have a "sleep" cycle of leaf movement: The leaves droop at night and rise in the daylight. If a silk tree plant (species *Albizzia julibrissin*, pictured in Figure **3.1**) is kept next to the hamster's cage, it too will show a shift when the lighting schedule is changed. At first it opens its leaflet pairs about 6 a.m. and closes them about 6 p.m. When the lighting schedule is changed to 9 a.m.–9 p.m., the plant gradually comes to open and close its leaflet pairs three hours later.

Figure 3.1 ***Albizzia julibrissin*** **(silk tree plant)**

(You can try this experiment yourself. It is described more fully in Chapter 7 of this book.)

Such shifts in timing can be more extreme. If the lighting schedule is shifted by 12 hours, both the animal and the plant will eventually become completely reversed in their activity cycles, beginning the day when they would normally be ending it and vice versa. People who work the "night shift" for a few weeks or more often experience this kind of complete reversal of the activity/ rest cycle.

Day Length

The environment more subtly influences the seasonal cycle of flowering plants. Plants flower at more or less predictable times each year, mainly in response to the number of hours of light during the day. But the relationship of flowering plants to light cannot be described as a simple response. Some plants form flowers when the days lengthen in early spring, some not until the longer days of summer, and others not until the days shorten in fall.

Other Sources of Environmental Influence

In the normal environment, it is usually difficult to investigate the effects of temperature changes on biological cycles. Temperature changes are usually accompanied by light changes—air and water temperature tend to increase during the day and during the summer. But in conditions of continuous light or dark, small temperature variations can coordinate some biological cycles. It is as if organisms have "backup" systems. If the main timing clue (light) is not available, then other timing clues can be used instead. For many cold-blooded organisms, a temperature change of five degrees is enough to affect biological cycles, and in some cases as little as one degree will do. In contrast, warm-blooded animals are much less influenced by environmental temperature changes.

Meal timing has a strong effect on some cycles, too, and may even overpower the effects of the light schedule. Various experiments have also suggested that some animals might use changes in air pressure, electric fields, or magnetic fields as timing clues.

The fruit fly (*D. pseudoobscura*) needs lighting and temperature stimuli to activate internal rhythms before it can emerge from the egg. For eggs kept in total darkness, the stimulus may be as brief as a 0.002 second flash of light. It is not clear whether the stimulation initiates the internal rhythms or merely synchronizes cycles which are already at work. In either case, the fact that a brief flash of light, carrying no other information, can stimulate complex rhythms is strong evidence for an endogenous clock.

Saunders, David S. (1977). Circadian rhythms: The endogenous oscillator. *Tertiary level biology: An introduction to biological rhythms* (pp. 19–39)

Evidence Against Control by Environment

There are whole books on the fine points of environmental influence—how the strength of a light or temperature stimulus can make a difference in when the peaks occur, how the comparative length of light and dark can affect the amplitude of a cycle, and so on. There are also many theories about what goes on inside an organism that lets it respond to the environment. But environmental influence is not the whole story.

Delayed Response to Schedule Shift

Changes in timing of cycles usually do not immediately follow shifts in environmental schedules. If a hamster or a bean plant has its lighting schedule changed by six hours, it may take as much as a week for its activity cycle to shift completely to the new schedule.

Figure 3.2 shows the results of an experiment on sugar storage in the livers of over 600 mice over nine days. (The plotted points are given as a percentage of the average measurement for that day.) The solid

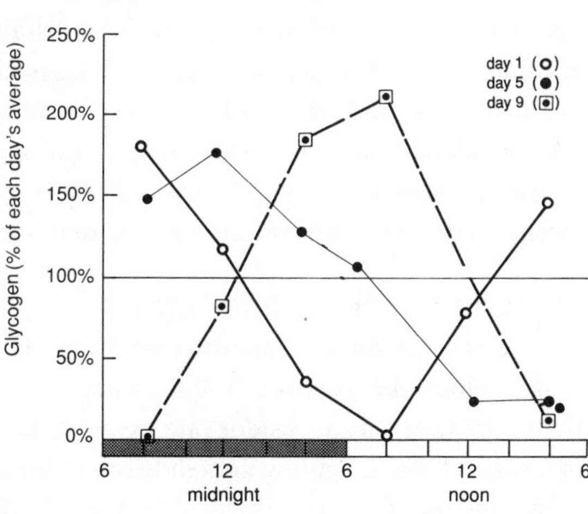

Figure 3.2 Adjustment of mice-liver cycle over nine days after a schedule shift of 12 hours

line shows the average amount of glycogen (stored sugar) on day one, when the mice were all still on a schedule of lights-on from 6 a.m. to 6 p.m. On the next day, the lighting schedule was shifted 12 hours—to lights-on from 6 p.m. to 6 a.m. On days five through nine, measurements were again made on samples of 20 mice every four hours. The cycle had shifted only about halfway to the new schedule by day five. It was not until day nine that the cycle had shifted the full 12 hours.

You may have experienced such a delay in adjustment yourself if you have ever had to start getting up several hours earlier than usual. You may have felt a little "off" for as much as a week while your body's cycles were shifting their peak times to

Recent studies have shown that daily exposure to very bright light for several hours can shift people's rhythms, as much as six hours in seven days. Some scientists believe this discovery could lead to cures for jet lag, some kinds of insomnia, and problems caused by changing work shifts.

Czeisler, C.A., et al. (1986, August). Bright light resets the human circadian pacemaker independent of the timing of the sleep-wake cycle. *Science*, 233, 667–670.

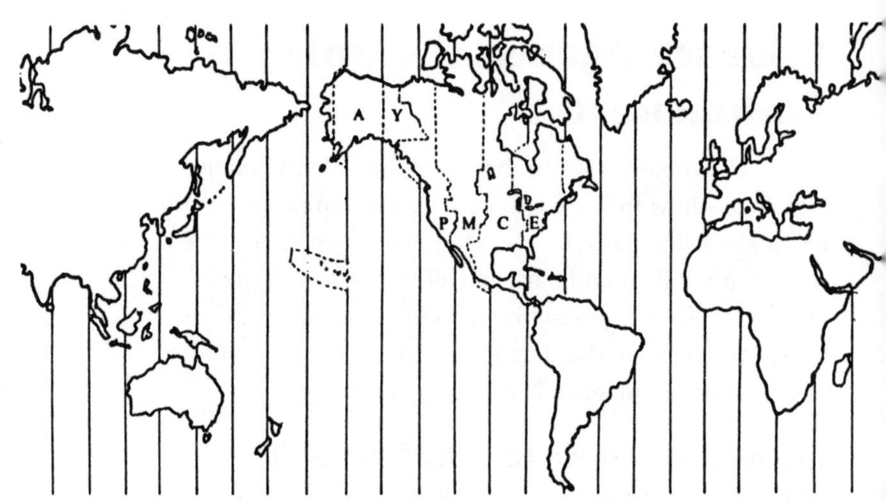

Figure 3.3 World time zones

match your new sleeping and eating schedule. Because people notice this discomfort after making long-distance flights to different time zones, it is called "jet lag." Jet lag is not the result of flying or of distance traveled; it occurs because your body needs time to adjust to a shifted sleeping and eating schedule when you arrive in a new time zone. (Figure **3.3** shows the time zones of the world, each different from the next by one hour.)

Direction of Schedule Shift

More evidence that the environment has only partial control over most biological cycles is that it is easier to shift cycles one way than the other way. Consider one experiment in which the results of 6-hour shifts in lighting schedule on a bird and a man were compared. For the first six days, man and bird were kept on a schedule of 12 hours of light and 12 hours of dark that was close to that of their usual environment. The times of sleeping were observed for both, and a record was also kept of the man's body temperature.

Figure **3.4** shows how the lighting schedule was shifted. In the figure, each row is a day, and the shaded area for each day shows the hours of darkness. The open circle in each row shows the time of waking on that day. (Notice that under normal conditions the bird wakes up before the light comes on, whereas the man awakens some time after.) The darkened circles for the man show when his temperature minimum occurred each day.

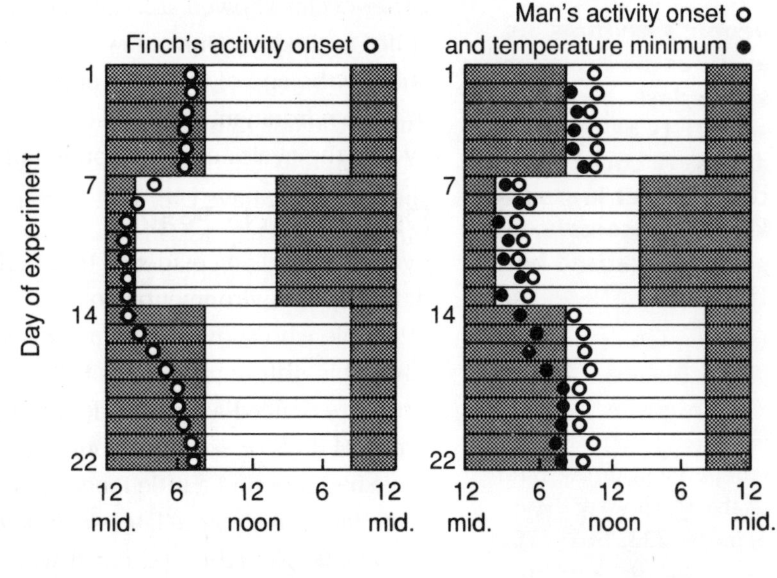

Figure 3.4 Comparison between the activity levels of a man and a finch before, during, and after a shift in light/dark schedule. As shown by changes in the shaded areas, the schedules were shifted six hours earlier on the seventh day and then returned to normal on the 14th day.

On day seven, the lights were turned off six hours early, and the new schedule was kept for seven more days. By studying the figure, you can see that the bird adjusted its waking time to this earlier schedule within a couple of days. The man shifted his waking time and circadian temperature cycle almost immediately.

On day 14, the lights were kept on six hours longer and the schedule was returned to the original timing. The man's hours of sleeping again adjusted quickly. However, the bird's waking time and the man's temperature cycle took four days or more to adjust completely to the schedule delay. It was plainly easier for some cycles to shift in the earlier direction than in the later direction.

Not all birds or all people show the same speed of adjustment to a new schedule. There are even differences in the speed with which different cycles shift in the same individual. In one experiment, a man's schedule was shifted 12 hours. For a week before the shift and for a week afterward, measurements were made every few hours on many biological variables. How would each variable respond to the reversed schedule? The peak in excretion of phosphate by the kidneys began to occur earlier, and in four days had made a full 12-hour shift ahead. The peak for calcium excretion also shifted to earlier and earlier hours, but in four days had moved only eight of the 12 hours. On the other hand, the peak for blood pressure began to occur later, but in four days had shifted only four of the 12 hours. Sex hormone excretion also shifted later, but in four days had shifted only two hours.

The 12-hour shift caused some of the man's biological cycles to shift earlier and some to shift later—and all at different rates. Other cycles showed still different rates of adjustment to the shifted schedule. Only a few had completed the 12-hour schedule shift by the end of the week of measurements. Although changes in the environment exerted strong influence on the biological cycles, the cycles each responded differently to that influence.

Drift of Cycle Peaks

Even more striking evidence that biological cycles are only partially under environmental control is found in a few medical cases of people who show about-daily cycles that peak progressively later. This drift of peak times is illustrated by the case of one man who experienced a cycle of depression that appeared to repeat every 24 days or so. It was found that his "daily" temperature cycle had a period a little longer than 24 hours, so his temperature peak time came a little later each day. If on one day his cycle peak was at 4:00 p.m., the next day it would be at 4:30 p.m., the next day at 5:00 p.m., and so on. Twenty-four days later his cycle peak would be at 4:00 a.m! After 24 more days it would be back to "normal" at 4:00 p.m., and so on. It is not surprising that he was disturbed by this repeated mismatch—every 24 days he would feel as if he were getting up in the middle of the night and going to bed in the middle of the day. His temperature cycle (and probably others, too) had escaped from the usual 24-hour environmental pattern.

You have seen several kinds of evidence that the environment has a strong influence on biological cycles: Changing the environment usually does cause changes in an organism's cycles. You have also seen several kinds of evidence that the environment's influence is limited: Biological cycles take a while to shift to new schedules, they are easier to shift in one direction than the other, different cycles shift at different rates, and some biological cycles may even escape the influence of the environment altogether. Because response to environment does not completely explain biological cycles, some additional explanation for biological cycles is needed. It seems that there is a lot of control coming from inside the organism.

Internal Sources of Rhythms

What would happen if there were no cycles in the environment—

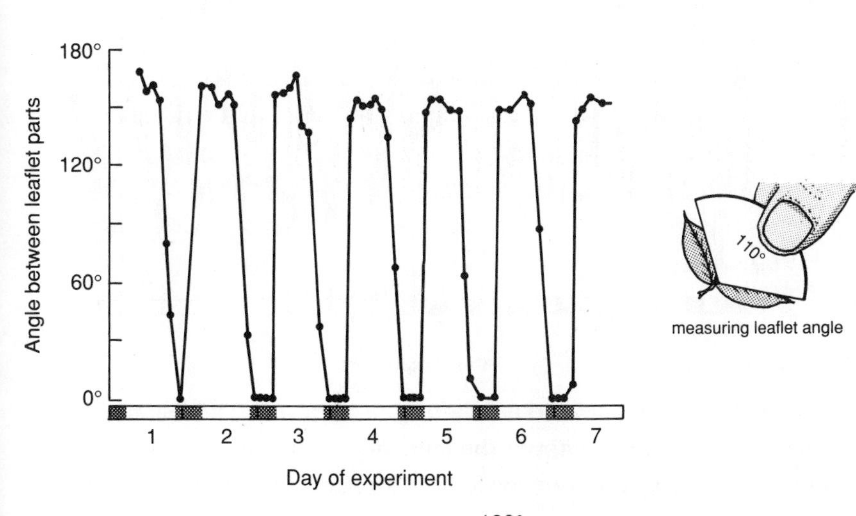

Figure 3.5 Leaflet angles for a silk tree plant on a schedule alternating 16 hours of light and eight hours of dark over one week

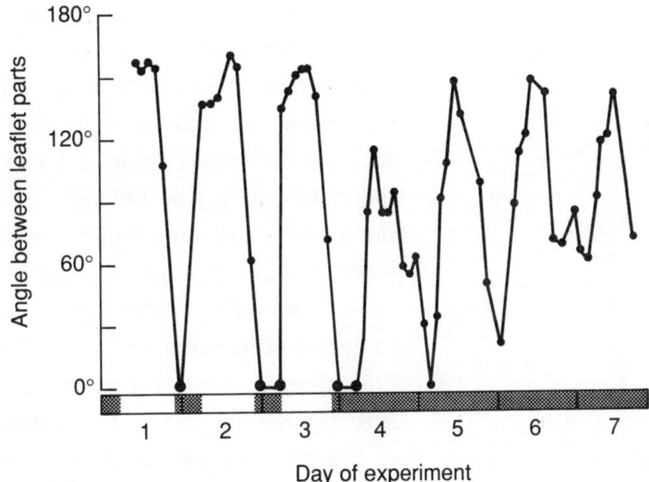

Figure 3.6 Leaflet angles of a silk tree plant kept on a 16 hours light/8 hours dark schedule, then kept in darkness for the next four days. The plant's leaf pattern continues without the influence of light.

that is, if environmental conditions were constant? What would an organism's internal system do by itself?

Persistence Under Constant Conditions

Many experiments have been done on plants, animals, and people to see what happens to cycles under unchanging conditions in a laboratory. Some plants, if kept in constant light and constant temperature, lose their leaf-motion cycle completely and immediately. In these plants the motion cycle seems to be controlled entirely by the environment. In other plants, the cycle will continue for a week or more in either constant light or constant darkness.

For example, Figure **3.5** shows a seven-day record of leaf angles for a silk tree plant kept on a repeating schedule of 16 hours of light and eight hours of darkness. (The dark sections on the time scale indicate periods of darkness.) Figure **3.6** shows a record for another silk tree plant kept on the same schedule at first, but

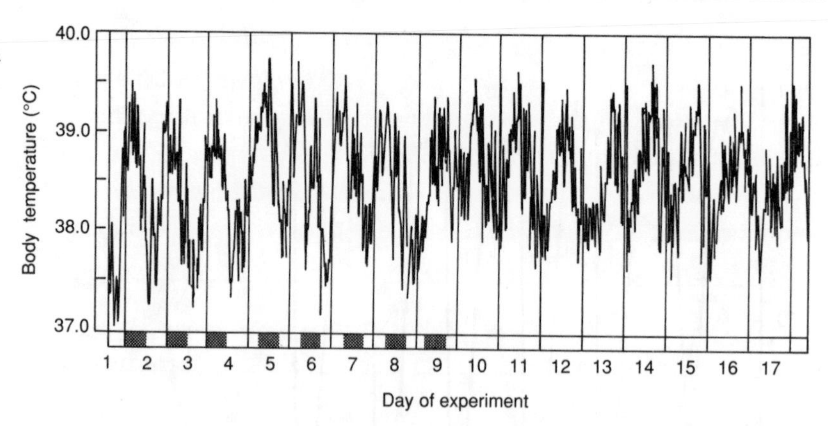

Figure 3.7 Internal temperatures of a female rat kept on a 12 hours light/12 hours dark schedule for nine days then kept in constant light. Under constant light conditions, the cycle persists.

then kept in constant darkness after the third day. The leaf movement is not as large without the influence of the light, but it does continue. The leaf motion cycle of this plant seems to be an innate biological cycle that is amplified by environmental influence.

This persistence of cycles without any known external influence from the environment may surprise you, but it has been known for over 250 years. In 1729 the French scientist Jean Jacques de Mairan published a description of how he had kept plants in the dark and observed the continuation of their "sleep movements." He concluded that the plant could "sense the sun without seeing it in any manner."

In animals, too, biological variables continue to rise and fall in cycles even when environmental conditions are kept constant. For example, Figure **3.7** is a 17-day record of the internal temperature of a rat. For the first nine days, the rat was kept on a schedule of 12 hours of light and 12 hours of darkness. From day ten on, it was kept in constant light. The temperature in the cage was kept constant throughout. (Measurements were made automatically: A tiny temperature-sensing radio transmitter was surgically implanted in the rat's abdomen, and the radio signals were detected by an antenna under the floor of the rat's cage.) There was only a slight reduction in the range of the cycle when the rat's lighting clues were taken away. Just as for the plants above, the cycle here was originating inside the organism itself. There also seems to be a four or five day cycle, (associated with the rat's estrus cycle) which is confirmed by statisical analysis. Less obvious but statistically demonstrable in longer series of data is an about-seven-day cycle.

An organism does not have to be as complicated as a silk tree plant or a rat to show the persistence of circadian cycles under constant conditions. Cycles persist even in one-celled plants and animals, even in simpler microorganisms such as the bacteria.

For instance, there is a species of one-celled algae (species *Gonyaulax polyedra*) that has a circadian cycle of glowing with a pale green light. Kept under constant conditions in a laboratory, this species of algae continues its glowing cycle indefinitely, as the results of one experiment show in Figure **3.8**. In the figure, the amount of light given off by a test tube of algae is graphed over a

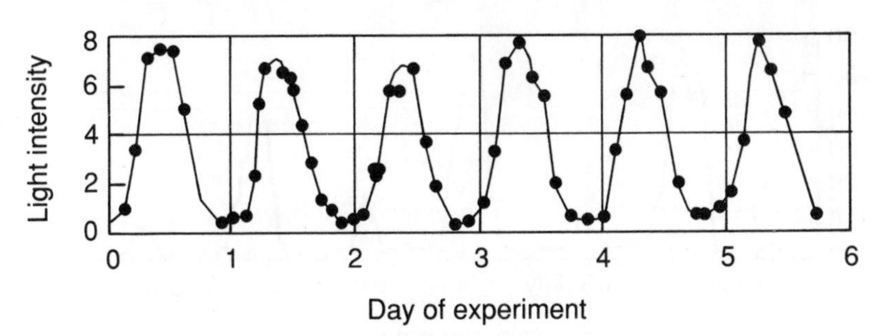

Figure 3.8 Intensity of glow from a sample of algae (*Gonyaulax polyedra*) under constant conditions over six days. Its glowing pattern persists.

span of six days. You can see that the period of the cycle is close to 24 hours.

About-Yearly Cycles

There are many examples of biological cycles with periods longer than a single day. These longer cycles are most often approximately a week, a month, or a year in duration. The hibernating squirrel experiment described in the Introduction is one example of persistence of about-yearly cycles. Such internal cycles are also found in migratory birds that spend their winters in the tropics. Experiments have shown that when these birds are kept under constant laboratory conditions, they show persisting about-yearly cycles in body weight, in molting, in size of male sex glands, and in "migration restlessness" (repeatedly hopping in the direction in which they would normally be migrating). Other experiments under constant conditions have shown that sheep maintain an about-yearly cycle in wool growth as do deer in antler growth.

About-Monthly Cycles

Crustaceans collected on sea coasts continue to show cycles of activity that match the tides where they were found, even when they are kept under constant laboratory conditions. An interesting example of the tidal cycle is found in a tiny crustacean called the sand hopper (species *Talitrus saltator*). This animal stays buried in the sand of the beach except when the edge of the high tide reaches it, allowing the sand hopper to swim for a while. In one experiment, sand hoppers were collected from the shore, placed in beakers, and kept under constant conditions in the laboratory for four days.

How do the sand hoppers in the laboratory know when to swim?

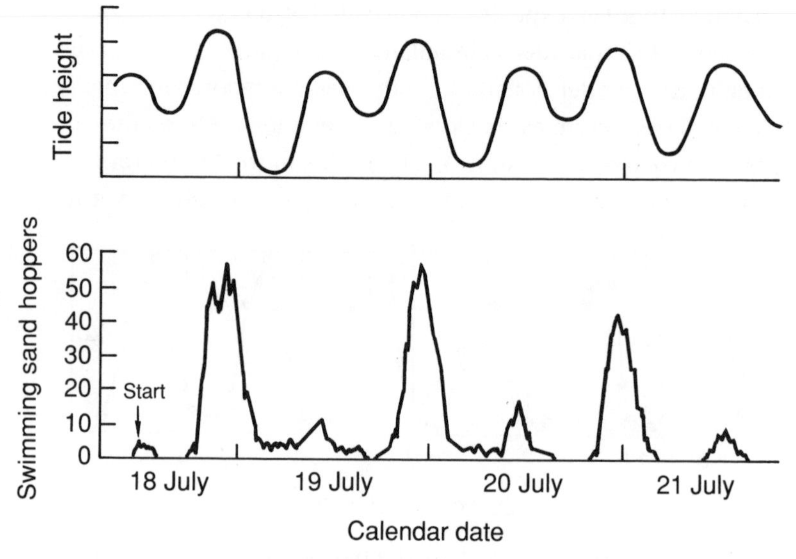

Figure 3.9 Comparison of sand hoppers' swimming activity in a laboratory beaker and the height of the tide where the sand hoppers had been collected. Even without the influence of the tide, the sand hoppers' swimming cycle still parallels the tide.

The lower graph in Figure **3.9** shows how many sand hoppers were swimming, recorded every three minutes from July 18 to July 21. The upper graph shows how the height of the tide varied over those four days on the shore where the sand hoppers had been collected. (Because of the position of the sun and moon on these dates, the two tides each day were unequal.) Though there was no tide in the beakers, the swimming cycles of the hoppers still appeared, just as if they were still periodically getting wet on the shore.

Are these cycles really caused within the living organism or tissue? Or is the environment still exerting some influence that the experimenters have not been able to keep constant? Can it be that organisms can somehow "sense the sun" as de Mairan thought over two centuries ago? How else could a biological system keep nearly perfect time?

Variation in Free-Running Cycles

The first answer to these questions is that persisting cycles usually do not keep perfect time—they do not match environmental cycles exactly. We have said that organisms under constant conditions show persistence of about-daily, about-weekly, about-monthly, or about-yearly cycles. In fact, the *free-running* cycles of organisms under constant conditions are seldom exactly the same as the environmental cycles they follow under normal circumstances. Figure **3.10** shows the activity records of two different species of fruit bats kept under constant conditions. The graphs show measures of their activity during each half hour of a 10-day experiment. The graphs for the 10 days are stacked on top of one

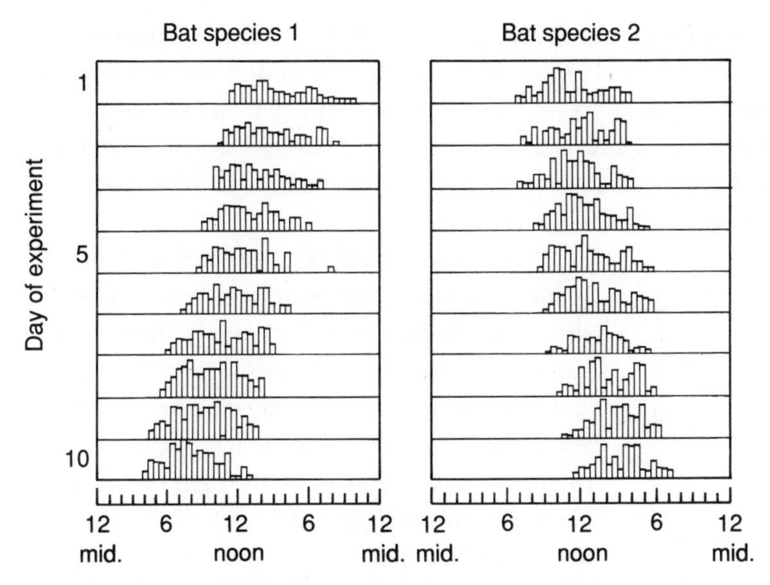

Figure 3.10 Activity of two species of fruit bats under constant conditions over ten days. Bat A shows a cycle period of about 23 1/2 hours while bat B shows a cycle period of about 24 1/2 hours.

another, with the first day at the top. Bat 1 started its activity about a half hour earlier each day, which means that its free-running period was shorter than 24 hours—about 23 1/2 hours. Bat 2 started its activity about a half hour later each day, so its free-running period was about 24 1/2 hours.

Such differences in free-running periods are found even among members of the same species. Figure **3.11** shows records of wheel-running activity for two flying squirrels kept separately under constant conditions in a 26-day experiment. Each squirrel had a steady drift in the time of day when it started to run con-

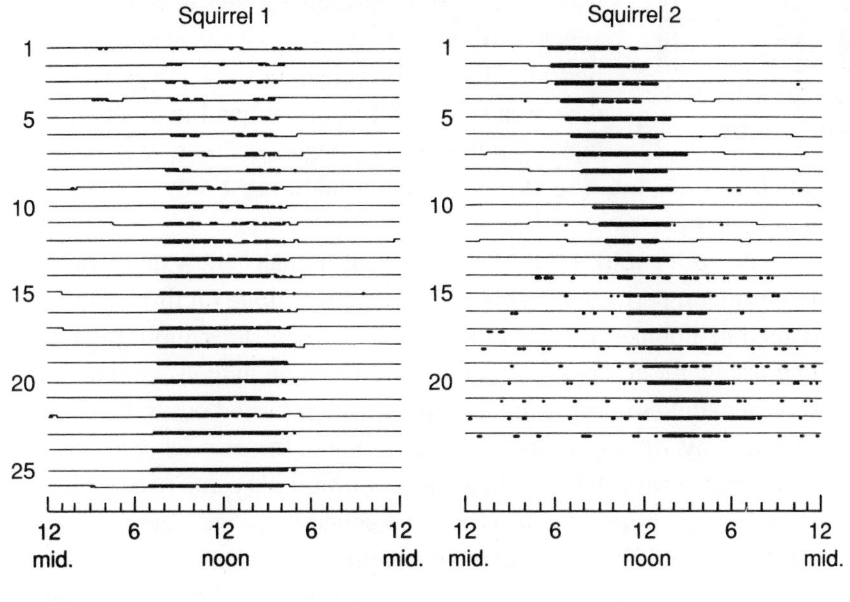

Figure 3.11 Activity records of two squirrels of the same species over 26 days. Their cycles also differ in period. Each mark in the figure indicates a turn of the squirrel's running-wheel. When wheel running is continual, the marks for each turn blur into a solid band.

Figure 3.12 Waking pattern of a human subject over 30 days; for the first six days, there were 16 hours of light and eight hours of dark; the next 18 days were in continuous light; and in the final six days, the schedule was returned to 16 hours light/8 hours dark.

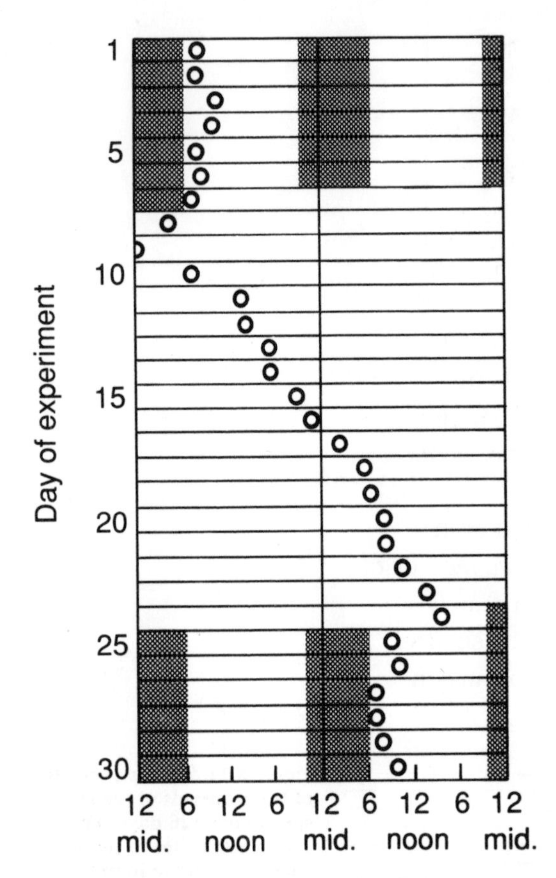

tinuously on its wheel. Squirrel 1's period was only two minutes short of exactly 24 hours, while squirrel 2's period was 21 minutes longer than 24 hours. In an experiment with 50 flying squirrels kept separately in constant darkness, all but five had free-running periods between 23 1/2 and 24 hours. This kind of variety in periods is typical of free-running cycles.

People who have been isolated for long spans of time in caves or in isolation rooms also show such drifts. Figure **3.12** shows the waking pattern of a human subject who spent 30 days in an isolation room. The only clue the man had about the time of day was the lighting in the room. For the first six days, the lighting schedule was 16 hours light and 8 hours dark. Then the researchers left the lights on continuously for 18 days before returning to the original light/dark schedule for the final six days. The circle in the row for each day shows when the man awoke. (The time scale at the bottom is two days wide, because he drifted so much that he would go right off the edge of a 24-hour-wide graph.) Under conditions of 16 hours light and eight hours dark, he kept fairly constant sleeping times. When time clues were removed, he woke earlier for a day or two, and then began to wake and retire progressively later each day. Inspection of the data shows that the average period of his sleep cycle over the free-running days was about 26.1 hours. When the original lighting schedule was begun again on day 24, he quickly locked into the old routine—but he thought only 23 days had gone by.

As with squirrels and bats, different humans show different free-running periods. An example is the study done on three men placed in a bunker for one month. They had no clues about time of day from light, temperature, or clocks. The men initially agreed on a schedule workable to all of them, which turned out to be a 27-hour day. After one week, however, one of the men felt very uncomfortable with the schedule and went onto his own. Two weeks later, the two other men broke off the schedule as well and went on their own daily routines. After a month, researchers opened the door of the bunker to find the three men seated at the table, eating together. Were the men back on a common schedule? Not at all:

One man was eating breakfast, one was eating lunch, and the third was eating dinner!

It is also common to find that different cycles in the same person have different free-running periods. For example, one human subject kept under constant conditions in a bunker settled into a sleep-waking cycle of about 33 hours. For four weeks his temperature was measured continually, and his urine was analyzed to find the amount of calcium and potassium excreted. Calcium excretion showed the same 33-hour period as sleep, but the temperature and potassium cycles showed a more typical period of 24.7 hours.

Environmental Synchronization

Clearly, most biological cycles are not totally controlled by environmental cycles. Under normal conditions a biological cycle may have the same period as an environmental cycle, but it will probably continue to run by itself when environmental conditions are held constant. In such free-running cycles, periods can vary among species, among individuals of the same species, and even among different systems in the same individual. These periods can also be affected by disease, drugs, or by social interaction among free-running individuals.

Many, if not most, of the biological cycles we observe in nature depend on chemical cycles that arise within organisms. We will call these cycles that arise from within biological *rhythms*. Under normal conditions, the environmental changes influence these rhythms, continually adjusting their peak times to keep them in step. Like the beat of a parade drum, the environmental cycles serve to keep "in step" cycles that are capable of "marching" by themselves. We say that internal rhythms are usually *synchronized* by environmental influences.

Limits of Synchronization

The environment, however, has only limited influence over internal biological rhythms. Although circadian cycles in some plants can be squeezed by artificial lighting schedules into "days" as short as four hours, most processes coordinated by internal circadian cycles cannot be forced into "days" shorter than about 18 hours or longer than about 36 hours. If an organism is given a lighting schedule with a period that is drastically different from its internal rhythm, the organism breaks free from the synchronizing control and goes back to its usual free-running period, usually not more than three hours different from 24 hours. Hamsters, for example, have a narrow range—their days cannot be compressed or stretched beyond 21–26 hours.

In summary, many biological rhythms arise from within organisms, although they are typically influenced by the environment in period, amplitude, and peak time. Might there be a single master source for timing information *within* organisms? In the next chapter we will look for the biological origin of rhythms and try to determine whether organisms contain a single "clock" for the coordination of all their internal cycles.

Study Questions

1. Do plants and animals respond immediately or gradually to changes in lighting schedules? Give some examples to illustrate your answers. How would it help or hurt an organism to respond to lighting changes?

2. Describe two examples of biological variables that show cyclical changes when the environmental conditions are kept constant.

3. Name animals that show the following:

a. a tidal cycle

b. a wool-growth cycle

c. a migration cycle

d. an antler-growth cycle

e. a hibernation cycle

4. Do all cycles in an individual run at the same rate? What evidence can you cite for your answer?

CHAPTER FOUR

Discovering the Wheel: The Origin of Cycles

Internal Timekeeping

We have discussed what biological cycles are, what their graphs look like, how we know whether they exist, and what environmental factors may influence their timing. Now we'll take a look inside organisms to see how internal biological rhythms are coordinated from within. In this section we will also consider the origin of rhythms and discuss how internal rhythms may have adapted to the rhythms of the environment for improved survival and reproductive success.

In Chapter Three, evidence was presented that although the environment influences cycles, a lot of the coordination of cycles comes from inside the organism. How does that control work? Can an organism create a complicated system of cycles all by itself? Actually, all that is necessary for any system to develop cycles is for part of the "output" of some part of the system to be fed back into it as "input." A familiar example is the squeal that comes from a sound amplifying system when there is too much "feedback." When too much sound output from the speaker reaches the microphone, the sound is amplified, gets back to the microphone, is amplified still further, and so on. The result is a large oscillation in the sound system, which produces the squeal—thus causing you to adjust the sound system.

Chemicals

There are many chemical-reaction systems that have this kind of feedback. For example, a chemical produced in a reaction can affect the rate at which the reaction itself proceeds, either speeding it up or slowing it down. When there are several steps in a reaction, the products of any one step may affect other steps, so there are many possible feedback effects. It is not surprising, therefore, to find biochemical reactions that speed up and slow down in cycles. The reaction with six chemicals described at the beginning of this book is one example.

Cells

The simplest of cells contain hundreds of chemicals that undergo complex reactions. If cyclical chemical reactions are possible with only six chemicals, we shouldn't be surprised to find that cells have internal cycles. (Remember the example of the glowing cycle in algae, p.28.) In fact, cyclic changes are the very basis of cell life.

Imagine a factory that makes cars. Different operations have to be going on at the same time and at different rates. For example, wheels have to be produced four times as rapidly as bodies, and screws have to be produced at a greater rate still. And products of operations have to be brought together at just the right time. Wheels cannot be put on before the axles; hoses and wires cannot be connected until the engine is in place; and so on. In the living cell, all of the chemical "parts" have to be ready in the right amounts and at the right time for metabolism, growth, and reproduction. This sort of activity is more than just a collection of interacting cycles; it is a system of synchronized cycles that has been shaped by evolution to provide efficient operation of life processes.

Tissues and Organs

Left to themselves, single cells show their own free-running cycles. But each cell is also sensitive to timing information from its environment. When cells are packed together to form a tissue, they influence one another. And the timing of all cells in a tissue is influenced by variations in the fluids that bathe the cells. Cycles in blood concentrations of nutrients and hormones can therefore have a synchronizing influence on the tissues of different organs.

However, that doesn't necessarily mean that all cells in an organ end up with the same timing. Among the many different kinds of cells in the pancreas, for example, there are at least three types whose division rates are distinctly different in timing, even though the cells may be side by side. It is not surprising, therefore, that the rhythms of cell activity in different organs are not the same. For example, most cell division in a mouse's ear occurs in the middle of the mouse's rest span; most cell division in a mouse's adrenal gland occurs eight hours later, when the mouse is awake.

Whole Body Synchronization by Hormones

Left to themselves, cells, tissues, and organs have their own rhythms. But they are seldom left to themselves. In a many-celled organism, some parts produce substances that are carried throughout the body and keep cell rhythms coordinated. In horseshoe crabs, for example, rhythms are coordinated by substances produced in the eyestalks. In some insects, hormones produced rhythmically in parts of the brain have been shown to synchronize other body rhythms.

An amazing demonstration of such coordination was shown in an operation on the brains of two species of silkworm moth. Moths of one species usually emerge as adults from their pupae at dawn. Moths of the other species usually emerge at dusk. In an experiment designed to show characteristics of the moths' biological rhythms, researchers first removed the brains of both species of moth. As a result, the moths emerged at random times across the 24-hour time span. The researchers then took each pupa's brain and transplanted it into the same pupa's abdomen. This time the original pattern of emergence was restored—one species emerged at dawn and the other species emerged at dusk. Finally, the researchers removed the brains of one species and transplanted them into the abdomens of the other species (and vice versa). The moths with transplants emerged at the usual time for the *other* species. Because the transplanted brains were not connected to the pupa's nervous systems, any timing effects they caused must have been due to chemicals they produced.

In birds, a gland in the brain called the pineal gland coordinates many body cycles. Removal of the pineal gland in birds seems to cause loss of rhythms in temperature and activity. There is also evidence that the pineal gland may play a role in about-yearly changes in the pattern of circadian rhythms. The pineal gland, however, does not run entirely by itself. It receives cyclic input from other parts of the bird's brain.

A Master Timer?

In mammals, many rhythms are influenced by the rhythmic release of hormones from the adrenal glands, which sit on top of the kidneys. This rhythm arises spontaneously within the adrenal gland. Any small piece of the adrenal gland shows a persistent circadian rhythm even when removed from an animal and kept in a glass dish. If the adrenal glands are completely removed from an animal, many of its other body rhythms disappear.

Can it be that the adrenal glands are the master timing device for circadian rhythms in mammals? The answer is no, for two reasons. First, not all rhythms disappear when the adrenal glands are removed. Second, the adrenal gland itself is synchronized by hormones from another organ—the pituitary gland.

The pituitary gland is connected by a short stalk to the

bottom of the brain, as shown in Figures **4.1** and **4.2**. This gland produces a variety of hormones which have many effects on body systems. Among the pituitary hormones is one with a strong effect on the adrenal glands, called adrenocorticotropic hormone, or ACTH for short. (Adrenocorticotropic means "hormone that affects the outer layer of the adrenal gland.")There is a distinct circadian rhythm in the amount of ACTH the pituitary secretes into the blood. The change in amount of ACTH in the blood synchronizes the spontaneous circadian rhythm of the adrenal glands.

You may wonder now whether the pituitary is a master timing device for circadian rhythms. Again, the answer is no. The pituitary rhythms are affected by many things, including the amount of *adrenal* hormones in the blood—another example of feedback! The rhythmic release of ACTH from the pituitary is particularly affected by hormones and nerve signals from a region of the brain connected to the pituitary—the hypothalamus. The hypothalamus region is cross-hatched in the larger diagram of the brain shown in the figure.

Is the hypothalamus the master source of timing information? As you might expect, the answer is once again no. Rhythmic activity of the hypothalamus is affected by many things, including the amounts of adrenal and pituitary hormones in the blood (more feedback). The hypothalamus also has input from the brain and sense organs.

Of special interest in the hypothalamus is a pair of cell clusters lying just above where the optic nerves cross. Destroying

Figure 4.1 Some influences on the production of an adrenal hormone

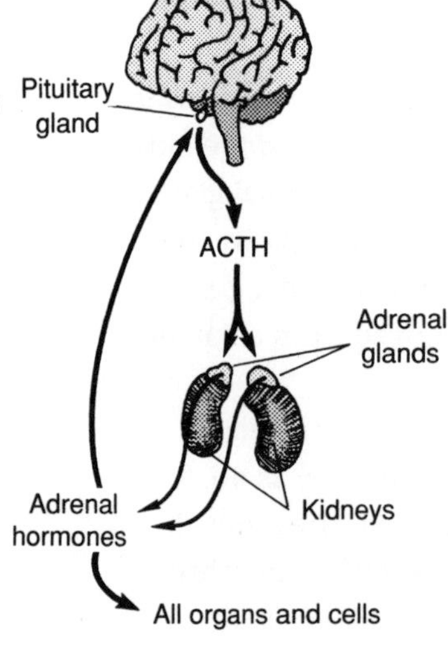

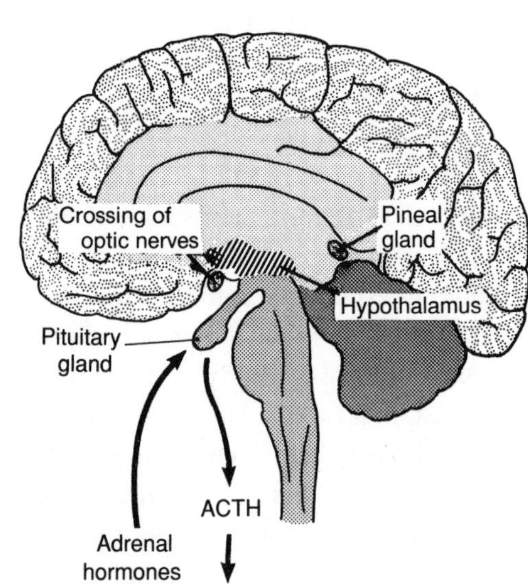

just these tiny regions of the brain has disruptive effects on many body rhythms. When the experiment was first tried on rats, scientists who saw the data concluded that the rats' body rhythms had stopped completely. A more careful statistical analysis of the data, however, showed that some of the rhythms had only been greatly reduced in amplitude and shifted in peak time. These cell clusters appear to be important for coordinating rhythms, but they do not cause the rhythms in the first place.

Ovulation Cycles

Because more is known about circadian rhythms than about longer or shorter rhythms, we have looked mostly at circadian rhythms. There is, however, a great deal known about one particular rhythm with a period of about four weeks: the menstrual cycle in human females (Figure **4.2**). This cycle, which involves almost every body system in some way, has a feedback system involving cells, hormones, and nerves.

How does the cycle work? An egg cell developing in the ovary will suddenly burst out and start its journey toward the uterus. The burst is related to an increase in the amount of LH, luteinizing hormone, a hormone secreted by the pituitary gland. This sudden increase in pituitary output is related, in turn, to an increase in the amount of luteinizing releasing factor (LRF) from the hypothalamus of the brain. What causes the increase of the LRF? The hypothalamus produces this releasing factor when the amount of still another hormone, estrogen, increases rapidly in the blood stream. What causes a rapid rise of estrogen in the blood

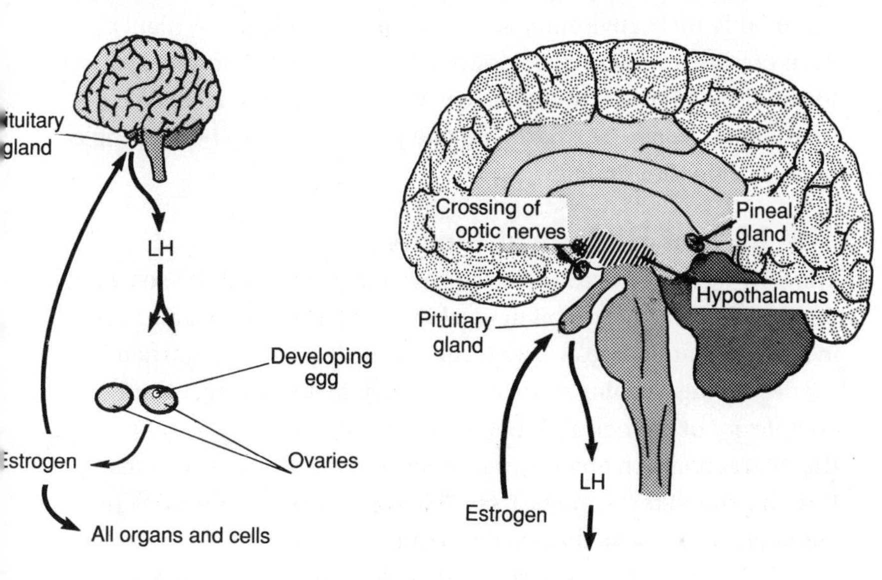

Figure 4.2 Some influences on the production of an ovulation hormone

stream? Perhaps you have come to expect feedback effects by now and have guessed—estrogen is produced by the developing egg cell in the ovary!

So here is another example of a rhythmic feedback system, one that involves influence on the ovary by the pituitary, the pituitary by the hypothalamus, and the hypothalamus by the ovary. (Several other hormones are involved, and despite many years of study the whole system is still not completely understood.) The control of the cycle seems not to lie in any one part, but in the interactions of the overall system.

Because there are so many cross-influences, it is not surprising that the menstrual cycle may sometimes be disrupted by unusual body conditions, such as the great amount of exercise of competing athletes. And, because the hypothalamus receives input from many parts of the brain, emotional upsets also may disturb the cycle. Remember also that cycles can have periods of seconds, months, or even years. These cycles affect one another, producing a vast web of interactions among different body systems.

These interrelationships look complicated, but the details are not the point here. What is important to understand is that biological rhythms arise from a complex system of influence and feedback from cells, chemicals in the blood, and nerve signals. The timing of cell activities is influenced by hormones from glands, which can be influenced by other hormones, which are influenced by parts of the central nervous system, which are influenced by hormones and signals from other parts of the central nervous system, which are influenced by sensory input from the environment. Because people can modify their environment (as by turning on lights at night), even our environment is influenced by our central nervous systems. Although the system has key parts that coordinate the information from other parts, no one part appears to be the single source of rhythm timing.

The Idea of Biological Clocks

You may have wondered why we have not mentioned the so-called biological clock, a term often used in popular magazines and even in some scientific books. The term "clock" is a useful shorthand for describing a timing system, but it may make us forget the complexity of biological timing systems and how they arise from the interaction of many organs throughout the body. (If you use this shorthand term, at least say "biological clocks" and think of a community of clocks that influence one another.)

Before we knew how complex biological timing is, many

researchers did try to locate *the* clock (a single master source of timing information) in test animals. They proceeded by removing parts, cutting connections, shocking, starving, and poisoning to block chemical processes. Often these assaults greatly disturbed, suppressed, or concealed rhythms by blocking the behavior that was used as a measure of the rhythm's presence. But if the organism survived, so did some patterns of rhythms. If an attack on an organism was enough to truly stop its rhythms, it was also enough to kill the organism. Knowing as you do now how rhythms are built into all life processes, you would expect that.

There is another reason why we have avoided speaking of a clock or even a community of interacting clocks. In looking for an ultimate source of timing for all biological rhythms, there may be a tendency to overlook the great practical importance of the rhythms themselves, regardless of what causes them. In Chapter Five you will see how important a knowledge of rhythms can be.

Tuning Internal and External Rhythms

Some textbook discussions of biological rhythms suggest that plants and animals developed them over time, in response to cycles in the environment. (In such a view, for example, there might have been prehistoric rodents that had no rhythms.) But, considering how vital rhythms are to life at all levels, it seems more likely that biological processes were internally rhythmic from the very first cells—from the beginning of life itself.

Internal and External Evolution

This does not mean, however, that rhythms have not evolved. If an organism has some characteristic that gives it an advantage in surviving and reproducing, it is likely to produce more offspring than others of its kind will. If the advantage is inheritable, these offspring will also be more likely to have offspring, and so on. This natural selection of advantageous characteristics leads, over the long run, to the evolution of organisms that function well in their environments. Rhythm characteristics can have two kinds of survival advantage: The internal system of the organism may function more efficiently, and the organism may operate more successfully in its external environment.

Rhythm timing is an inheritable characteristic. Varieties of fruit flies and yeasts have been found with distinctly different free-running rhythm periods that are inherited. The period is apparently affected by a single gene that can mutate, giving an organism a 19-hour or 21-hour period instead of the normal 24-hour period—or giving it no obvious rhythm at all. As animals have become more complex—with organs, circulatory systems, hormones, and nervous systems—rhythms, too, have had to grow in complexity to keep the whole internal system running smoothly. Once an organism has evolved a smoothly running system of internal rhythms, however, it must also be successful in its relationship with the *external* environment.

Animals that have some natural advantage in fitting into the environment are more likely to survive than animals that do not. As has already been implied, one such natural advantage is a match between an organism's internal rhythms and the rhythms of the environment. There are at least four ways in which having such matched rhythms can aid an organism in its survival and reproductive success: anticipation, efficiency, competition, and navigation.

Anticipation means going somewhere (or preparing for something) a little ahead of time. With a time sense, an animal can arrive early enough to find food or mates, and can leave in time to avoid predators or unpleasant weather. A bird, for example, would not do too well if it had to depend on the sun to wake up in the morning, or depend on colder temperatures to leave for warmer climates. An "early" bird that can anticipate the sunrise can, as the old saying goes, "catch the worm." Rhythms for animals lend many of the same advantages that people get from wearing wristwatches and keeping calendars.

Efficiency is the ability to achieve the most benefit with the least expenditure of energy. For example, if an animal is likely to encounter a meal during only one part of the day, then there is little need for its food-processing system to be kept ready around the clock. Rhythms of alertness and strength peak during an animal's most active hours (daytime for squirrels, nighttime for owls), and can be lower during off-hours. Rhythms for the repair of fast-wearing tissues, such as skin, peak during the hours of rest.

For a common example of efficiency through timing look at how a thermostat can be used to keep a room optimally warm. An ordinary thermostat turns the heat on whenever the room temperature falls below a set point—and turns the heat off when the temperature rises above that point. Heating is most efficient if the thermostat is adjusted for different times of day. When people are at home and active, a higher set point is used than when they are away from their homes. Likewise, people sleeping in bed can stay warm with blankets, allowing a lower set point and a reduction in fuel consumption.

You could, of course, change the setting of the thermostat by hand when you went to bed at night and got up in the morning. But the heating system takes a while to warm the house. A thermostat with an automatic timer, however, will turn on before you get up, so the house will be warm when you want it to be. (Notice that this example of efficiency also in-volves some anticipation.)

The coordination of human body temperature appears to use such a sophisticated thermostat with a rhythmic change of setting. When your body starts to get too warm or too cool, the amount of blood flow to the skin or the amount of perspiration changes to keep your body temperature within a half degree or so of normal. But the "normal" temperature varies throughout the day with a circadian rhythm, ranging from a high of about 37.3°C (99.2°F) to a low of about 36.4°C (97.5°F). The rhythmic peak occurs late in the waking hours and the rhythmic low occurs late in sleep (about two hours before waking).

Competition for resources involves both *anticipation* and *efficiency*, but there is also an advantage to sharing time well with other species. Different species of flowers bloom at different times each day—and at different times of year. Some animals are awake only at night, others only during the day. Thus, not all animals are trying to get the same food and water sources at the same time, and not all plants are attracting insects for pollination at the same time. Even within the day and night "workshifts" there are early, middle, and late animals and plants.

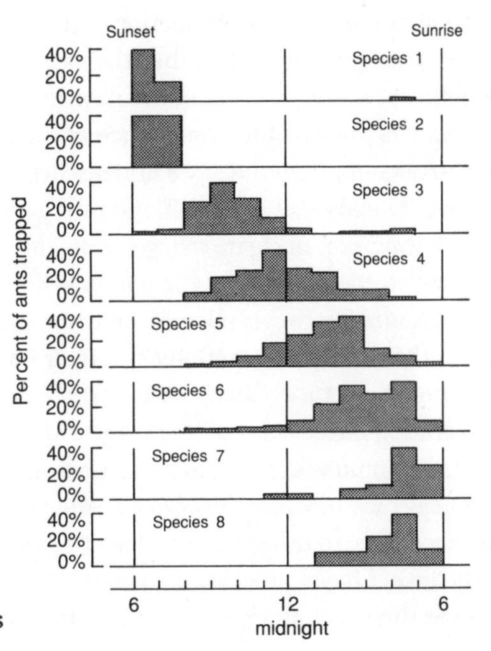

Figure 4.3 Different "work shifts" for eight species of ant

Figure **4.3**, an example of such "time territories," shows activity differences in eight closely related species of ants. The bars in the figure show the percentage of ants of each species that walked into traps at each hour of the night. Most of the ants of species 1 and 2 were captured between 6 p.m. and 8 p.m., while ants of species 8 were generally captured between 1 a.m. and 6 a.m. Why such a difference? If one ant species is actively foraging when other ant species are not, there will be less direct competition among the ants. There may even be different food sources available at different times of day.

Even within a species, there can be time territories as well as space territories. For example, a dominant animal in a troop can choose its own space, leaving the others to use whatever space is left; but it can also choose its own time to use limited resources (such as a water hole), leaving the others to do so in what time is left. Clearly there is an ecology of time, in which each species—even each individual—has its own way of fitting into the overall pattern of timing.

You can use polarized sunglasses to detect differences in light from different parts of a blue sky. Rotate your polarized lenses as you look around. The angle that makes the sky darkest will differ according to where you look relative to the sun.

Navigation is the process of relating where you are to where you want to be. Many species of animals, from flies to humans, use the sky for navigation. Some bees can use the sun for navigation, and some ants have a mechanism that allows them to sense subtle differences in sunlight scattered from different parts of the sky. Thus they can use a patch of blue sky to orient themselves even when they can't see the sun. Birds can navigate on clear, moonless nights by using the stars, as demonstrated by laboratory studies in which birds shown "fake" stars in shifted positions fly in correspondingly shifted directions.

The position of the sun or stars, however, can indicate a reliable direction only in connection with the time of day. For example, the sun may be in either the east or the west, depending on whether it is early or late in the day. So what enables an animal to navigate is not just the appearance of the sky, but the ability to take into account how the sky's appearance changes with time. There are animals that are unable to take account of time in this way and, thus, can navigate using the sky for short trips only, during which the position of the sun or stars does not change much. Presumably these animals have not yet evolved a link between their circadian rhythms and their navigation system.

Whole books have been written on the navigation of homing and migrating birds—and of tiny insects whose brains are no larger than pinpoints. (Bees in particular have fascinating methods of keeping track of where flowers are and communicating their experience later to other bees.) Almost all of these feats require that the insect have internal "information" about the time of day. Whatever the exact source of animals' time sense, experiments show that it can be readily shifted by changes in lighting schedule, and so seems to be connected to another biological rhythm.

Development of Rhythms in Individuals

Because even single cells have rhythms, it is not surprising that some rhythms in complicated organisms are evident very early in their development—even in the sprout or embryo stage. Mammals, for example, develop rhythms in temperature and blood pressure some time before birth. Other rhythms (menstruation for example) depend on cells that have not yet developed fully at birth. Even if all the "parts" are in place, some plants and animals have to experience certain stimuli from the environment before some of their rhythms appear. In humans, some rhythms do not appear until days, weeks, or months after birth. For example, Figure **4.4** shows how the pattern of wakefulness changed in a group of 19 children from age three weeks to six months. (In the first two weeks wakefulness appeared to be random, its occurrence being equally likely at any time around the clock.) During weeks three to six there already was a slight difference in night and day wakefulness. By weeks 23 to 26, the difference increased, resulting in an eight-hour sleep span similar to that of adults (for which parents are grateful).

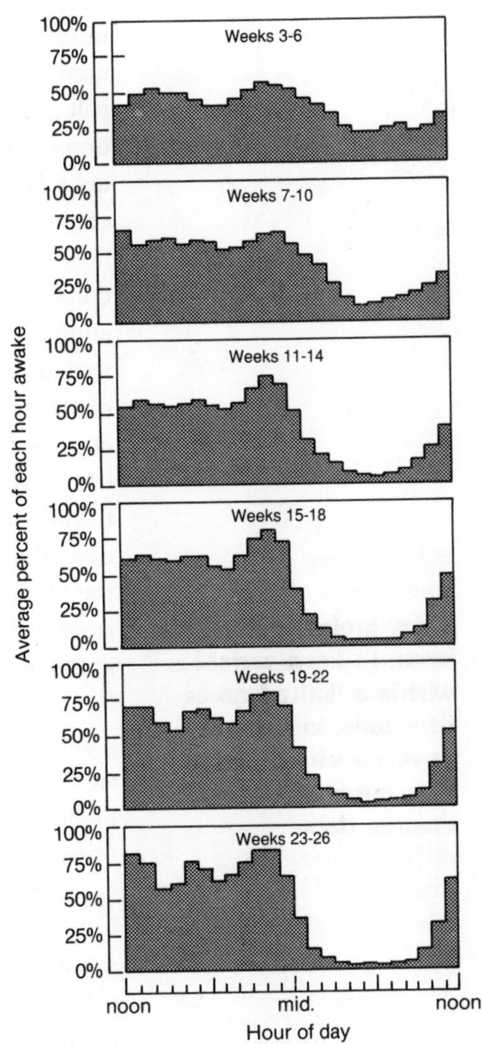

Figure 4.4 Wakefulness of human infants during each hour of the day at six different ages

Cycles in sex hormones, a longer-period human rhythm, do not appear for many more years—adolescence—when the individual is very conscious of the change. Much later there is the disappearance of the menstrual rhythm in women. There is also evidence that as some people grow old other rhythms may decrease in amplitude or may lose their pattern of synchronization. Thus our pattern of rhythms changes throughout life from before birth into old age, as part of the pattern of life, reproduction, and death. In the next chapter, we will explore how we can use knowledge of rhythms to make life better—and perhaps longer.

Constancy Versus Rhythmicity

For a long time, the scientific study of biological rhythms was held back by rigid belief in an old idea. This idea, known as homeostasis, says that the physiology of an organism always acts to keep internal conditions *constant*. For example, one of the most precise control systems in the human body regulates the concentration of salt (sodium chloride) in your blood. Whether you eat a lot of salt or as little salt as possible, several chemical systems in your body act to keep the concentration of sodium in your blood from changing by more than about two percent. Although limited in how much they can vary, internal conditions are not truly kept constant. Even with the extremely small change that is allowed in sodium concentration, there are distinct circadian rhythms in the variation. As was true for the body's tightly regulated temperature control system, there is a daily variation in the "set point" for controlling sodium concentration. And because some systems in the body are extremely sensitive to small changes in the concentration of sodium, even a small change can be important. Moreover, the narrow range of variation in sodium concentration in the blood is achieved through a large rhythmic variation in the amount of sodium excreted into the urine, thus maintaining strict limits on internal concentrations.

Many biological systems serve to keep variables within a limited range of values, in spite of environmental influences that might tend to change them.

There are other large rhythmic changes: The number of circulating white blood cells, for example, varies about 50 percent over the span of a day, and the amount of circulating adrenal hormones varies rhythmically by as much as 25 percent. Large or small, internal rhythms serve to maintain the functions of life.

Organisms try to maintain not *constant* internal conditions, but *optimal* internal conditions. Optimal means that every body system is kept operating at a level that relates best to other body systems and to the cycles of the environment as well. If some physiological variable goes beyond the upper or lower limits of the rhythm pattern, the organism may function poorly or even die. So optimal conditions for an organism are a stable pattern of rhythmic change within rhythmic limits. To use the familiar example, your body does not try to keep your internal temperature always within one degree above or below a value of 37.0°C (98.6°F), but tries to keep it always within a half degree above or below a value that varies rhythmically between 36.4°C (97.5°F) and 37.3°C (99.1°F).

The idea of homeostasis is largely true and was an important advance in biology. However, many scientists believed so strongly in this idea of internal constancy that they dismissed evidence of variation as unaccountable "noise" of no importance. Physicians typically look at whether medical measurements are within the "normal range." Such a range covers most of the variation observed in healthy individuals, but usually takes no account of when the measurements are made. Because we know now that a significant part of the variations is predictable, it is possible to figure out a more precise, rhythmic "normal range"—as discussed in the next chapter.

Study Questions

1. Give an example of a biological cycle influenced primarily by an organism's surrounding environment. Then give an example of a cycle influenced primarily from within the organism. Compare the survival advantages of the two types of cycles.

2. What are four ways in which matching between internal rhythms and environmental circumstances serve to help an organism to survive? Try to give examples different from those in the text.

3. Define "feedback" as it applies to biological rhythms.

4. In multicellular organisms, what is the general name of the circulating substances that act to synchronize internal cycles?

5. Is there a biological clock? Explain your answer.

6. Infants seem to cry at all hours; however, when they reach approximately six months of age, they appear to follow a sleep/wake pattern. How could you account for this noticeable change in behavior?

7. How does the idea of biological rhythms differ from the idea of *homeostasis* (the tendency of an organism to maintain a set of constant conditions inside its body)?

8. It is possible that the same value of a biological measurement would be interpreted as unusually high at one time and unusually low at another time. For example, how would you interpret an oral temperature of 98.0°F(36.8°C) in the graph below?

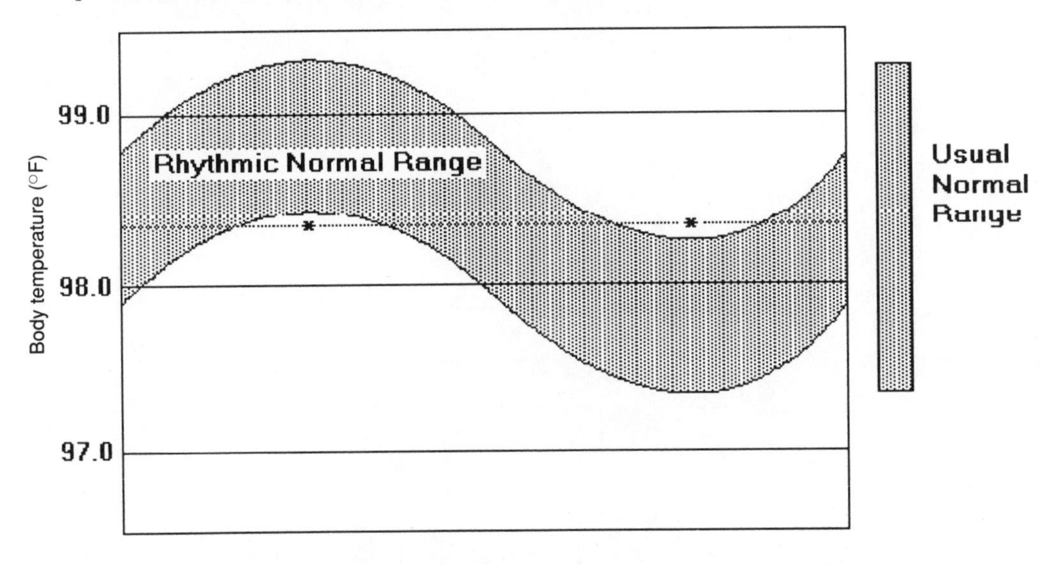

CHAPTER FIVE

Keeping in Step with Rhythms

Shifting Schedules

Studying biological rhythms is important for reasons other than the gathering of abstract scientific knowledge. While a better understanding of the life functions of organisms is important for its own sake, there are also some extremely practical applications of the study of chronobiology. As you will see in this chapter, a knowledge of biological rhythms can help us detect and treat disease, increase productivity in the work place, and improve our quality of life in a variety of ways.

Because biological rhythms involve complex internal timing systems, organisms cannot adjust immediately to new environmental schedules. If you change your sleeping schedule, say by getting up several hours earlier, your various body rhythms may take from a few days to a few weeks to shift completely to the new schedule. While you are adjusting, your various body systems will not be "up" when you would normally expect them to be. Furthermore, because the different rhythms are adjusting at different rates, the rhythms will not relate to each other in their usual pattern.

When people travel east or west to different time zones, they usually change their sleeping and eating behavior to match the new time zone. Unless they travel slowly, their bodies will not keep up with the schedule change. The outside mismatch (between your body and the clock on the wall) and the inside mismatch (between the different rhythms in your body) cause you to do things less effectively than you usually do. This is important to know in any situation where it is imperative that you do well—school examinations, business deals, diplomatic negotiations, surgery, or athletic competition. If your performance is important, you should allow plenty of time for your body to adjust before you perform.

An example of how a schedule shift hurts performance is

shown in Figure **5.1**. For nine days before a trip to Italy, seven healthy people in Minnesota took self-measurements every three hours on a coordination task. As you can see by the graph, their average performance improved steadily over the nine days—the time required to complete the task became shorter and shorter. Then the group flew by jet to Italy. You can see how the average performance got worse immediately after the trip. Performance began to improve as the days went by in Italy, and when they flew back to Minnesota after 20 days there was almost no loss of skill. (Remember that it is easier to shift in one direction than the other.)

A special problem of changing schedules is found in people required to do "shift work," in which sleeping and eating schedules may change as often as twice a week over many years. Because a complete shift of all rhythms to an eight-hour schedule change might require several weeks, workers who change shifts more often than that might *never* be in their normal, synchronized condition. Even staying permanently on a night shift often does not seem to help shift workers, because to be with friends and family, workers tend to go back to normal day schedules on their days off. That keeps their rhythms from synchronizing to the reversed night-shift schedule. Researchers are investigating to what extent production, safety, and health may suffer from such repeated changes. Since it probably is not possible to do away with shift work, the research often focuses on finding which shift patterns do the least harm.

Combating jet lag is a major concern for people who often travel long distances. In a new time zone biological cycles can take several days to regain their synchronization. While there is still no *cure* for jet lag, a drug called triazolam can help prepare a person for a new time schedule. Researchers also recommend that travellers begin a trip well rested, avoid caffeine, alcohol, and candy en route, and immediately assume the schedule of the new time zone upon arrival. Harvard chronobiologist, Charles Czeisler, thinks one day the interior lighting on long-distance flights may be adjusted to help reset passengers' biological clocks.

Chapel, R.J. (1985, November). The ravages of time. *Nation's Business, 73*(11), 92.; O'C. Hamilton, Joan. (1987, October 26). You don't have to give in to jet lag. *Business Week*, (3023), 126; Seidel, W.F., Roth, T., Roehrs, T., Zorick, F., and Dement, W.C. (1984, June 15). Treatment of a 12-hour shift of sleep schedule with benzodiazepines. *Science, 224*, 1262–1264.

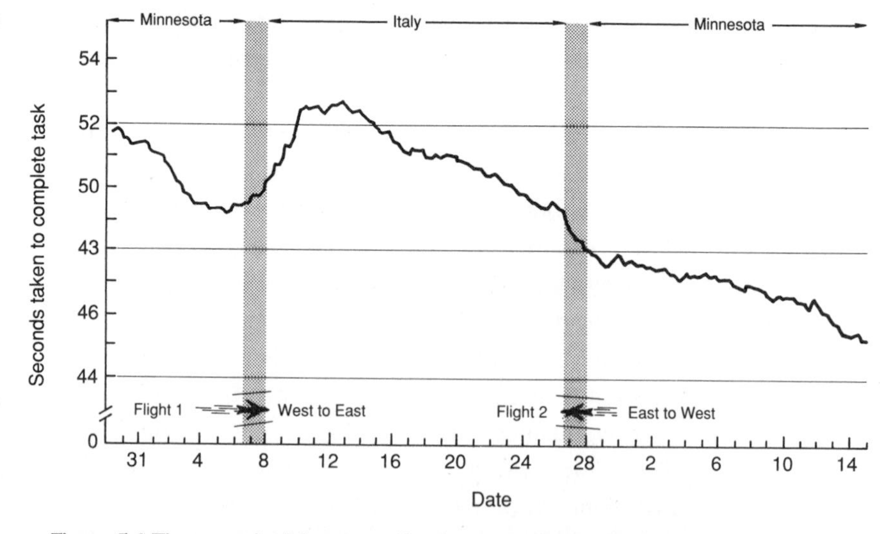

Figure 5.1 Time required for a coordination task obtained before, between, and after two intercontinental schedule shifts. Improvement of performance first declined but then resumed after a week on the new schedule. The opposite shift upon return to the original schedule in the U.S. had little effect.

Some workers, including many military personnel and some nurses, are on continually rotating shifts. One common shift-work requirement is two days of morning shift, then two days of afternoon shift, then two days of night shift, then three days off. (By now you should recognize this as likely to have complex effects on rhythms.) Figure **5.2** shows the results of a study made on people following this shift pattern. The graph shows the average variation in temperature for the group (dashed lines). The graph also shows the average performance on a test of crossing out certain letters on a page of type—a stand-in test for ability to attend to details. The solid circles show the average search speed when only two letters were to be crossed out. The performance peaks at about the same time temperature does. (Alertness tasks often show this relationship to temperature.) However, when there were six letters to be remembered, looked for, and crossed out, the worst performance came at the time of the temperature peak. (Memory tasks are commonly found to peak at different times than alertness tasks.)

Clearly it is not possible to give one time when someone is at his or her *best*. Depending on the task or the combination of tasks involved, one's best time could occur anywhere around the clock. Studies of the best shift patterns for workers must take into account the kind of work that is being done.

In our clock-governed society, most people run their lives on similar standard schedules, regardless of the subtle differences

Chronobiology is becoming an important part of business for many American companies. Work-schedule consultants are being hired by utility companies, overnight delivery companies, retail chains, and law firms. For a fee, the consultant analyzes the work to be done and the current shift schedules, and then prescribes a schedule. The analysis is lengthy and expensive, but at least one consultant guarantees that the new schedule will earn back the client's investment within two years.

Siwolop, S., Therrien, L., Oneal, M., and Ivey, M. (1986, December 8). Helping workers stay awake at the switch: Scientists are showing companies how to translate body rhythms into saner, more productive shifts. *Business Week*, (2976), 108.

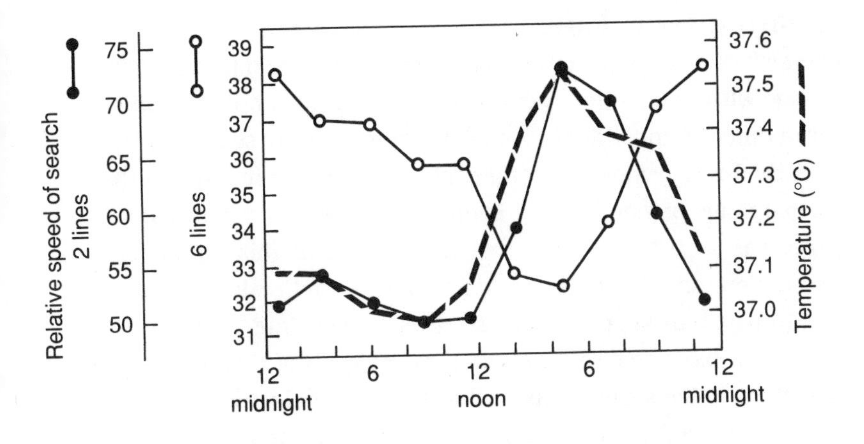

Figure 5.2 Relationship of temperature cycle for shift workers and their performance on two different visual-search tasks. Skill in spotting two letters parallels temperature, whereas skill in spotting six letters does just the opposite.

in the way their bodies run. A greater knowledge of rhythms could give businesses, hospitals; and even schools a choice of schedules—allowing people to choose the one that suits them best.

A 1987 study suggests that of junior and senior high school students divided into three groups, morning alert, afternoon alert, and evening alert, those who are morning alert consistently get the highest grades. A similar study indicates that morning-alert teachers receive better performance evaluations than teachers whose alertness and efficiency peak during the afternoon or evening. Can you suggest an explanation? Several alternative explanations?

Chance, P. (1987, October). The early bird makes the grade. *Psychology Today, 21*(10), 22.

Finding the Rhythm Range

Much research in biology is concerned with what things are *usually* like. In what habitat is an oriole usually found? What is the usual size of the pituitary gland? What is the usual number of white cells in the blood? The answers to questions like these make up a large part of what you read in science texts. They are the *facts* about nature—what things are observed to be usually like.

Variation Among Individuals

Collecting facts is easiest if things are always the same—if an oriole is always in about the same habitat, the pituitary gland is always the same size, the white cell count is always about the same, and so on. As you know well by now, however, there is a great deal of variation in biological facts. This makes research much more difficult. You cannot look at only one organism, or in only one field location. You must look at many organisms and many locations and report *ranges* of findings. For example, the normal range of nesting sites for northern orioles is from the tops of low bushes to the lower branches of tall trees; the normal size of human pituitary glands ranges from 0.8 to 1.5 cm in diameter; different healthy people have different white cell counts; and so on. Imagine a researcher recording one figure for the average height of the girls in your class for use in a scientific study. You might question the usefulness of this study since it fails to acknowledge the *range* of heights of the girls in your class.

Variation With Time

The task of describing nature becomes even more difficult if things change with time. If the size of a pituitary gland is not only different in different individuals but also varies with the season of the year, then researchers must take more measurements to find the normal range of gland size. It is understandable that not many biological "facts" have been investigated with all this care. It is also understandable that scientists have not yet realized the extent to which predictable variations exist over time.

Most biological researchers now realize that there are daily variations in life systems, as well as the seasonal variation that has been recognized for a long time. They recognize that measurements made at different times of day can yield different results, although many researchers still tend to consider the variations as purely random rather than partly rhythmic (and as inconvenient rather than potentially important). For some researchers, taking time into account only means reporting the time of day that treatments or measurements were performed. This information is of course not enough to detect rhythmic changes, but can it at least be helpful to other researchers who want to consider timing in interpreting the results? Not much, as you will see in the following section.

Time of Day Versus Time of Organism

You already know that "time of day" is much too crude an idea for biological research, because lighting and feeding schedules can affect the timing of biological rhythms. Yet reports can still be found in scientific journals stating, "the measurements were made at about 10 a.m.," or even just "in the morning," with no information at all about lighting or feeding schedules. If an experiment is to consider the effects of biological rhythms, the feeding, lighting, and temperature schedules must remain the same for all test subjects. Other outside stimuli may also be important, depending on the character of the study being conducted.

For a satisfactory experiment, test subjects must also be given time to adjust to the schedule that will be followed during the experiment. As you know, rhythms do not immediately change their timing to match a new schedule. A researcher should not only report the characteristics of the experimental schedule, but also how long the organisms had been kept on that schedule before the experiment itself began. Readers can then judge whether this was long enough for the organism to synchronize all its rhythms with the new schedule. Experimenters are completely safe from the confusing effects of rhythms only if they study the rhythms as part of the research.

Finding Differences

Biological research does not just describe averages or ranges; it also describes *differences* between organisms. Does a group of animals given one kind of food do better than another group given a different kind of food? Does a field treated with a new insecticide have more dead creatures than a field where it isn't used? Does a new drug X have more effect than the old drug Y? If the effects of different treatments depend in part on the rhythms of the organisms, then the results of the experiments will also vary depending on when treatments are given or when measurements are made.

In one study, for example, a group of treated mice was found to have white cell counts that were higher, the same, or lower than an untreated control group, depending on the time of day the measurements were made. [See Chapter 6] How would the results have been reported if the researcher had taken only one set of measurements? A researcher taking measurements early in the day might believe that the treatment *lowers* the white cell count. Another researcher, taking measurements later in the day, might publish results claiming that the same treatment *raises* the white cell count.

So far this chapter has focused on the problems rhythms can cause in scientific research. There is a positive side, however. Once these problems are understood, taking account of rhythmic variation can open new doors in biological measurement. A knowledge of rhythms, when put to use in the laboratory, can sharpen our interpretation of measurements taken at a single time. Further, rhythms themselves offer a new subject for research—what they are, how they can affect scientific experiments, and how they are changed by experiments. If life is fundamentally rhythmic, then our understanding of all fields of biology will be improved by studying the complex role rhythms play.

Detecting Disease

Knowledge of rhythms is essential to all stages of health care, from defining health to detecting risk, from diagnosing disease to giving treatment. The most obvious use of this knowledge may be in defining what is "abnormal" for a healthy individual.

Normal Ranges

For some biological measurements it is especially important to know the normal value as precisely as possible. This is most obviously so for measurements that relate to health and disease—body temperature, blood pressure, blood chemistry, and so on. Healthy values for these measurements vary from person to person. They are also different within the same person at different times. If the variations with time are unpredictable, then there has to be more uncertainty in the normal range. But if the variation with time *is* predictable, the measurements can be interpreted more accurately.

Because white cell counts have rhythmic changes of as much as 50 percent; because concentrations of blood hormones have rhythmic changes of as much as 80 percent; and because blood pressure readings have rhythmic changes of as much as 20 percent, etc., it is vital that standards for what is healthy take account of those rhythms. Adrenal hormones make a good example. Some people have a disease in which their adrenal glands often produce too much, and other people have a disease in which their adrenal glands often produce too little. But even in healthy people, the normal rhythm in hormone level swings from being close to the "too much" level in the morning and close to the "too little" level in the evening. To tell the difference between someone with an abnormal condition and someone with a normal rhythm low, it is necessary to measure at a time when the usual level is high. On the other hand, to tell the difference between someone whose level is always too high and someone who only peaks to a high level rhythmically, one must measure when a normal rhythm would be low.

Judgment of whether your temperature or blood pressure is above normal of course depends on what you believe normal to be. For example, would an oral temperature of 38.0°C in the early morning indicate a fever? What about in the evening?

Rhythm Characteristics

A knowledge of rhythms, however, can do more than improve the interpretation of single measurements. What if, for example, a disease affected not only single measurements, but a person's entire rhythm pattern? If we take a series of measurements at different times, we can describe what a person's normal rhythms are like. Then disease may be diagnosed by observing changes in the nature of a person's rhythms.

For example, let's look again at the temperature record shown in Figure **2.2**. There we showed only a five-day record, but in fact it was part of a nine-day record, shown in Figure **5.3**. Notice

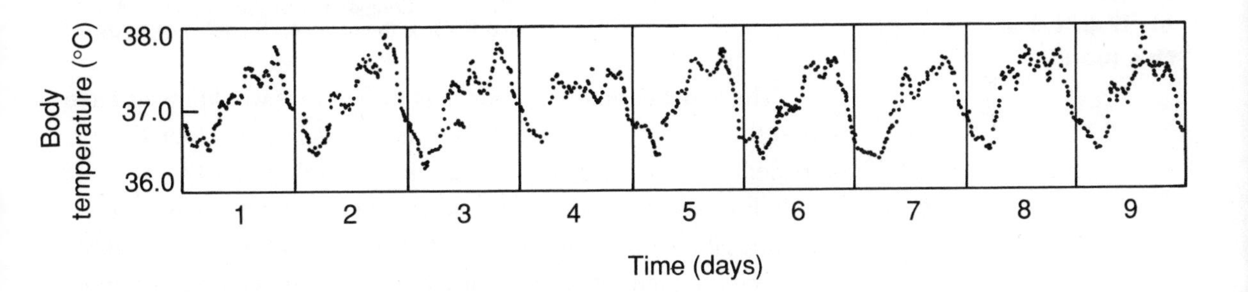

Figure 5.3 Man's body temperature recorded automatically over nine days

that the man's temperature is seldom measured at exactly 37.0°C (98.6°F), which is widely believed to be normal. About half the time it is higher, and half the time it is lower. Although his *average* temperature is close to 37.0°C, a measurement of exactly 37.0°C would be unusual at any time except during brief periods in the morning and evening.

Figure **5.4** shows the temperature data for each of the 9 days plotted within one 24-hour span. The gray area has been drawn so that it just covers all the daily graphs except day four. The dashed line shows how the record for day four compares to the hour-by-hour range for the other eight days. It is very close to the gray area except for about three hours during the afternoon. Was something wrong with the man on day four? There was. He had been diagnosed as having laryngitis on that day.

Figure 5.4 Comparison for recorded values for day 4 and the full range of values for the other eight days (from Fig. 5.3)

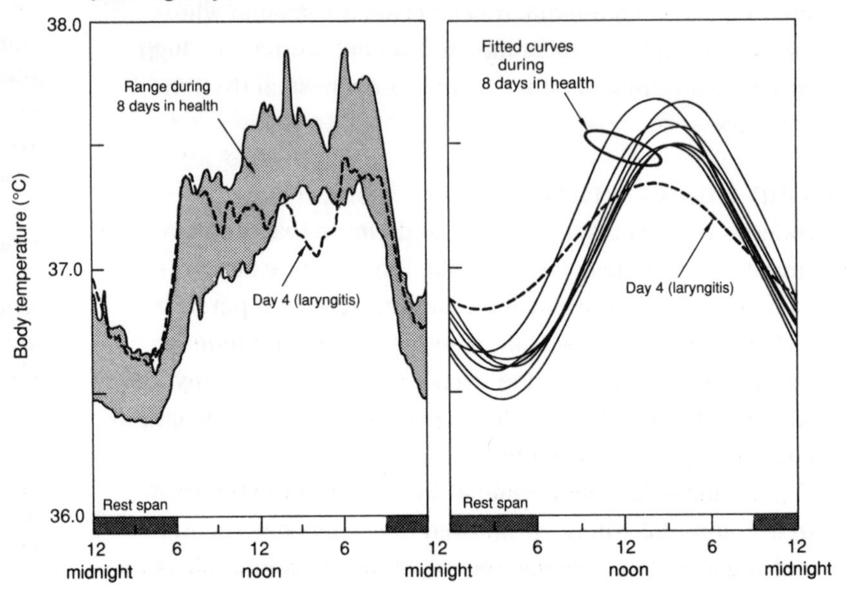

Figure 5.5 Comparison of sine curve for day 4 and the best-fit curves for the other eight days

How would you interpret a temperature of 37.0°C for this man?

The usual definition of normal temperature would never have led to the suspicion that anything was unusual on day four. Even knowledge of the normal *range* for this man (about 36.4°C–37.9°C) would not have led to the suspicion that something was wrong on day four. His temperature was low for him only at this time of day.

The uniqueness of day four can be seen better when smooth curves are fitted to each day's data. Figure **5.5** shows the curves for the eight days of health (solid lines) and day four (dashed line). What is different about day four is not that the average temperature is higher or lower than usual, but that the *amplitude* of the daily rhythm is smaller.

Fitting the data with a simple curve is not just a way of displaying information more neatly; it also can be a more sensitive way of testing for abnormality. Even if we leave out the measurements from the three afternoon hours when the day four record was noticeably lower than usual, the fitted curve for day four is still significantly different from the fitted curves for the other days.

This example shows how the description of the general characteristics of the underlying rhythm may be more helpful in some cases than the detailed description of the minute-by-minute data.

There is evidence that cancers disrupt the normal rhythm of the tissue in which they grow. One study of breast cancer shows how rhythm analysis might be used to detect disease much earlier than would otherwise be possible. Automatic records of temperature on the left and right breasts of one patient showed that the rhythms on the healthy breast had distinct periods of about 24 hours, about seven days, and about four weeks. The record for the other breast (in which a tumor was found) showed a loss of all three of these rhythms and the appearance of a 21-hour rhythm instead. Had the diseased condition of the one breast caused the circadian temperature rhythm to lose its synchronization with the 24-hour environment schedule and free run with a different period?

Analysis service is now available. If you make a series of measurements with an automatic blood pressure device, you can have your data analyzed and compared to rhythm "norms" for people of your sex and age. Write for instructions to: University of Minnesota Chronobiology Laboratories (5-187 Lyon Laboratories Minneapolis, MN 55455).

In the future, rhythm analysis might make it possible to detect high risks of disease before other symptoms of the disease actually appear. There is some evidence, for example, that children who are prone to developing high blood pressure may show unusual blood-pressure rhythms many years before they show generally high levels of blood pressure—perhaps even as newborns. Early detection might make preventive treatment possible, but once-a-year measurements will not do. Taking blood pressure data over several days and analyzing it for rhythm characteristics, perhaps as part of a school project, has the potential to reduce the high toll of high blood pressure.

As part of the research in this area, at least one college offers—through its health service department—an automatic temperature-recording device and a free computer analysis of data. The future may see high school biology students taking automatic measurements of a variety of variables and having a computer analysis of the rhythms recorded on a permanent health record. Every few years a new series of automatic measurements could be made and rhythm analysis done, perhaps by automatic telephone connections to a medical center or perhaps by a home computer. The new rhythm characteristics would be compared to the previous ones and any unusual changes would lead to follow-up study by medical professionals. Such a system could ensure better detection of disease, or even of potential disease, than our present one-measurement physical exam every few years.

Treating Disease

The study of rhythms can aid in treating disease as well as in diagnosis. The body's response to various medical treatments—surgery, drugs, X rays, heat, exercise, etc. depends on when they are used. Timing treatments so that they fit in optimally with the body's rhythmic activity can aid in a patient's recovery.

Surgery

Except in emergencies, most surgery is performed on weekday mornings, with no attention to the biological rhythms of either the patient or the medical staff. Their biological rhythms, however, are crucial factors in the success or failure of surgery. If patients' and surgeons' rhythms could be carefully timed, the risks in non-emergency surgery could be greatly reduced. Patients would be as resistant to shock and infection as they can be, and surgeons would be as alert and well-coordinated as they can be. (Some surgeons actually do go against hospital routine and schedule surgery late in the evening when they feel they are at their best.) Sensitivity to anesthesia is the most critical variable in non-emergency surgery, and there is extensive evidence that biological rhythms play an important role here.

There is evidence that an important about-seven-day cycle exists in the human immune system. For example, transplanted kidneys are most likely to be rejected by patients' immune systems at multiples of

Figure 5.6 Number of threatened rejections of human kidney transplants, on successive days after the transplant operations. (Total of 170 rejections among 508 patients.) An about-seven-day rhythm in the immune system would explain the high incidence of rejection on days 7, 14, 21, and 28.

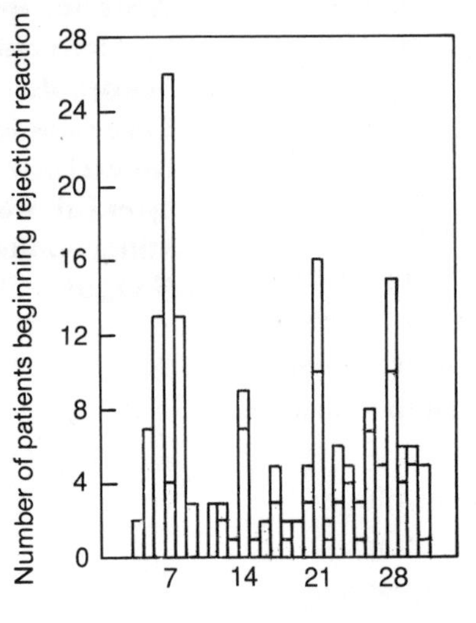

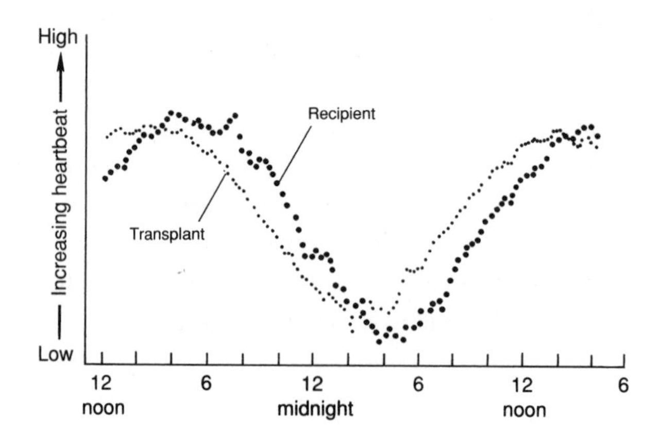

Figure 5.7 In a heart transplant patient, a comparison of the beating rates of the pacemaker section remaining from a patient's own heart and the beating rate of the transplanted heart; the two circadian cycles do not peak at the same time.

seven days. Figure **5.6** shows the day of threatened rejections of transplanted kidneys for 508 patients. As you can see, there are significant peaks of rejection at about 7 days, 14 days, 21 days, and 28 days after the operation.

Heart transplants could be affected as well. Figure **5.7** shows two beating-rate records for a patient 32 days after he had a heart transplant. The dark dots show the rhythm in beating rate for the pacemaker section remaining from his own heart, to which the transplanted heart was attached. The light dots show the rhythm for the transplant. They both have periods of 23.4 hours, but the transplant beating peak comes about two hours earlier than the patient's own pacemaker peak. Would the transplant have been as successful if the transplant heart were even less synchronized? Perhaps it will become important to take account of the patient's timing and the timing of the organ to be transplanted, and even to go to the trouble of shifting the recipient's schedule before the transplant operation.

Drugs

Because most body processes are rhythmic, it should not be surprising that the body will vary with time in its responses to drugs. For example, there have been studies of differences in the effects of alcohol on people's behavior depending on the time the drug is administered. Serious harm, however, is best demonstrated in studies on laboratory animals. Figure **5.8** shows the results of an experiment in which laboratory mice were injected with

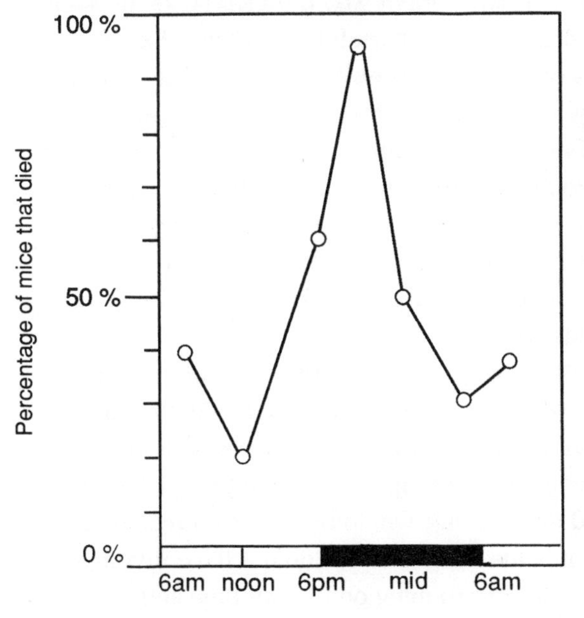

Figure 5.8 Percentage of mice killed by a single injection of alcohol at different times during light/dark schedule

Administering the same drug at different times is like administering different drugs. One study done on 31 patients with ovarian cancer concerned administration of the drugs adriamycin and cisplatin. One group of patients was given adriamycin in the morning and cisplatin 12 hours later, in the evening. The other group was given the same drugs, but with the times switched. After eight months, results showed that in the group receiving cisplatin in the morning and adriamycin in the evening, twice as many patients required dosage reductions and delays in treatment, four times as many treatments had to be delayed, and general treatment complications occurred twice as often as in the group which received adriamycin in the morning and cisplatin in the evening.

Hrushesky, W.J.M. (1985, April 5). Circadian timing of cancer chemotherapy. *Science, 228*, 73–75.

alcohol at different times. The horizontal axis is the time of day that each group of mice was injected, and the vertical axis is the percentage of the group that died from the alcohol. You can see that as few as 25 percent of the mice in a group might die at one time, and as many as 90 percent at another time.

Similar experiments have been done with many different drugs, in particular those that are used for treating cancer. To see the importance of such studies, you must first understand that cancer involves the uncontrolled growth of cells. Therefore, one way to fight cancer is to poison rapidly reproducing cells. But some normal cells reproduce rapidly, especially the hair follicles, the lining of the intestine, and the bone marrow cells that produce red and white blood cells. Chemical treatment of cancer (chemotherapy) tends to poison these cells as well as the cancer cells, and so causes intestinal upset, loss of hair, and very low blood cell counts. As unpleasant as the first two are, it is the drop in blood cells that is the most dangerous. Because white cells are a line of defense against infection, people receiving chemotherapy for cancer sometimes die of simple infections that a healthy immune system would routinely overcome. In a sense, the treatment may kill the patient before the cancer. Because of such deadly side effects, anti-cancer drugs must be given with great care.

If the rhythm of white blood cell reproduction is known, it becomes possible to give less anti-cancer drug when its side effects are likely to be worst (when cell reproduction is at its peak), and to give more when side effects will not be as bad (when cell reproduction is low). The patient would thus receive the same total amount of the drug with less harmful side effects. Experiments on thousands of mice have shown that such rhythmically timed treatment results in far fewer deaths from side effects and in greater cure rates. One series of studies showed that the cure rate for leukemia was doubled by timed treatment with the common anti-cancer drug, ara-C. When several drugs are used together, researchers found even more impressive improvements in cure rate. Similar studies involving human chemotherapy have begun, with promising results reported.

These important results have been obtained without yet considering what the rhythms of the cancer itself might be. Some cancers do have rhythms; their times are not necessarily the same as the white cell production rhythm, or a large improvement in cure rate from timed drug dosage would not be observed. In fact, there is evidence that some other cancers do not have rhythms: They act as if every cell were running on its own time and not responding to signals from other cells or from the body. Other cancers appear to have free-running rhythms with periods longer or shorter than 24 hours.

X-radiation is another means of "poisoning" rapidly reproducing cells. In one study of human cancers of the mouth (where it is fairly easy to take temperature measurements), it was found that the X-ray treatment at the time of peak tumor temperature was far more effective than treatment given eight hours earlier or eight hours later. When easily measured "marker" rhythms like temperature can be used, treatment may be adjusted for individual patients or even for individual tumors within patients. But much research is still needed before this marker-rhythm method can be commonly followed.

It may be possible to shift rhythms to give treatments the best possible chance of success. Using the cancer example again, imagine that the rhythm of a particular cancer closely matches the timing of the side-effect rhythm in the patient. Rhythmic timing of drugs would then have the same effect on both the cancer and the patient, with no gain in effectiveness of treatment. But if the patient's own rhythm could be shifted, possibly by changing his or her sleeping or eating schedule, then the two rhythms could be, at least temporarily, shifted apart. Drugs or X rays could then be used at just the right times to hit the cancer harder than the patient. Again, these possibilities are in the future, after much more research on laboratory animals and careful trials on people.

Timing is everything for the mayfly (*Dolania americana*), a sand-burrowing insect that lives near rivers and streams in the Southeastern United States. Although the life cycle of a mayfly lasts about two years, the adult stage accounts for only two hours (one 10,000th) of that time. It is during this period that mating takes place. What would happen if a mayfly emerged from its cocoon several days (or even several hours) late?

Peters, W.L. and Peters, J.G. (1988, May). The secret swarm. *Natural History, 97* (5), 8–14.

Rhythms in Ecology

Many of the same principles we have talked about in surgery and medicine can be applied to plants and animals that give us problems. The poisoning of undesirable insects or plants, for example, might be done with less poison if it can be applied at just the right time in the organism's rhythms. And by reducing the amount of poison required, we also reduce the adverse impact on other organisms—including ourselves.

Insect control has even been done without any poisons at all. In one experiment, female insects from one area of the country were transported to another area and released. Although the same species as the unwanted insects, the out-of-state insects had a different peak time for releasing the odor that attracts the male insects. The males, sensing the odor at the wrong time of day, became confused and were discouraged from mating at all. The result: fewer new insects.

Time relationships in ecology are an almost unexplored field of study, and you might be interested in studying them yourself. Both our understanding and our ability to get along with nature will be improved by knowledge of how organisms share time as well as space in an ecosystem.

Study Questions

1. List several jobs you know that involve a rotating shift. For each job, list an advantage and a disadvantage of such a schedule.

2. On a map of the continental United States, draw (approximately) the four time zones. Then describe, in your own words, what jet lag is. Can you get jet lag flying north or south? Can you get jet lag without flying at all?

3. You're planning to travel from Georgia to Oregon to enter a scholastic competition. Your plane will arrive the day before the event. What preparations can you make while you are still in Georgia to insure your best performance?

4. Define the phrase "normal range" for characteristics of an organism. How would "time-qualified" normal ranges be different? Would they be wider or narrower than ordinary "normal ranges?"

5. Explain the difference between "time of day" and "time of organism." What information about time should be reported in research on living organisms in order to be able to interpret the results?

6. How could knowledge of biological rhythms help kill insect pests more effectively and with less pollution of the environment?

CHAPTER SIX

Early Adventures in Chronobiology

A Puzzle of Fluctuation

This book on biological rhythms was written by Andrew Ahlgren, a science educator, with the collaboration of Franz Halberg, a pioneering researcher in medical biology and the man who named circadian rhythms and the science of chronobiology. At one point in their collaboration, Ahlgren asked Halberg how he had first become interested in rhythms. His answer was the basis for this chapter.

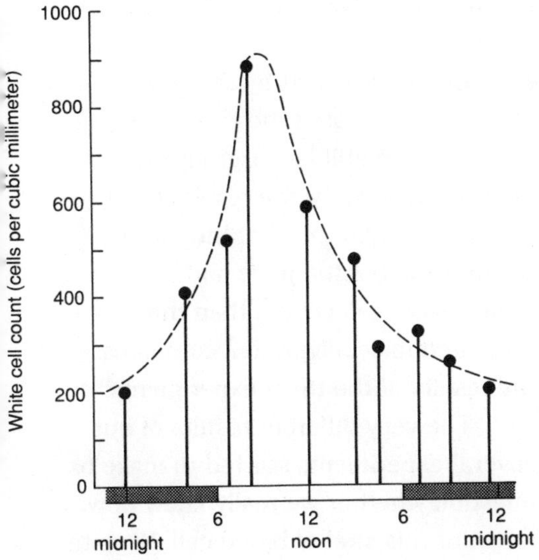

In the late 1940s our laboratory at Harvard University was investigating the effects of certain hormones in mice. One strong effect was on the number of a certain white blood cell type, eosinophils. But the results were often confusing because white cell counts varied so much. We suspected that one reason for the variation was the handling of the animals while taking blood samples. A frightened mouse might produce another hormone change that would reduce the number of these white cells. So we began to take special precautions to upset the animals as little as possible when taking blood samples.

To our surprise we found that when we reduced the amount of handling the cell count showed even more variation than it had before. Now, however, there was a predictable

Figure 6.1 Mouse white cell (eosinophil) counts made at different times during the light/dark schedule.

pattern. The average number of these white cells would drop from high counts in the morning to low counts in the evening, changing from approximately nine hundred cells per cubic millimeter to around two hundred cells. By reducing the irregular variations produced by the procedure, we had uncovered a regular underlying daily cycle. Figure **6.1** is a graph of how this white cell count of mice was found to vary with time of day. So the answer to the first puzzle was to take into account the time of day. Subsequently we found that changing the lighting schedule would shift the peak times. So time of animal was what counted, not simply time of day.

A Puzzle of Timing

In 1950 we began an experiment on the number of white cells in the blood of two groups of mice. (All of the white blood cells in this chapter are eosinophils.) One group of mice was undergoing experimental treatments, while an untreated control group was used for comparison. Because we knew by then that there were daily rhythms in white cells, we kept both groups on precisely the same schedule of dark and light, and we made all measurements in the morning. Measurements of the two groups indicated that, on the average, the treated animals had less than half the number of white cells as the untreated animals.

These early results were taken by the head of the department to a scientific conference in Paris. While he was on his way, we repeated the experiments on larger groups of animals, to be sure our findings were correct. This time, however, the two groups showed almost no difference in white cell counts.

Figure 6.2 Comparison of cell counts in control and treated mice in three experiments. The results of the three experiments were clearly different. The ratio of the two groups reversed from week one to week three.

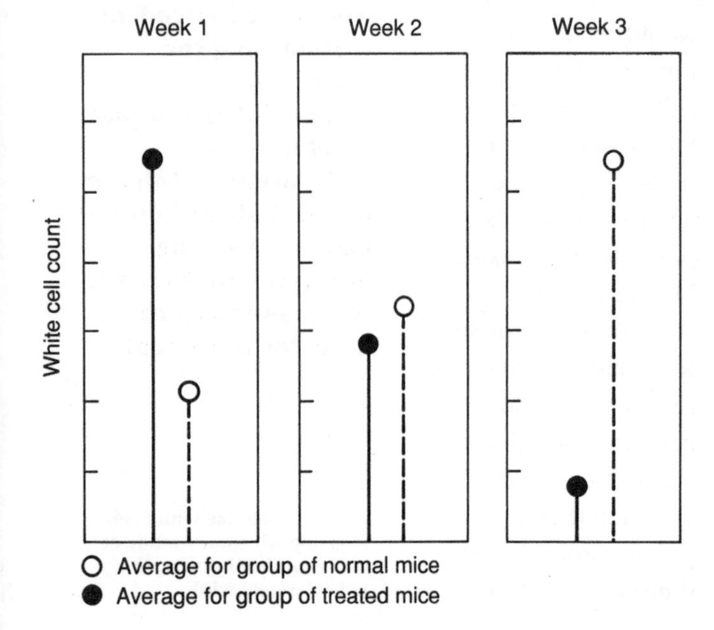

Greatly alarmed by the change in results, we repeated the experiment a week later on still larger groups of mice. Unbelievably, the results were just the opposite of what they were before—the treated mice had over six times *higher* cell counts than the untreated mice. Figure **6.2** compares the results of the three experiments.

The very different results of our several experiments started to make us question whether we really knew how to count this kind of blood cell. Despite our puzzlement, however, we recognized another possibility. As the number of animals in the experiment increased, it was necessary to begin making measurements earlier to complete them in the morning. Could there have been some unexpected results of making earlier measurements? Using the same mice as in the last experiment, we measured them again four hours after an early morning first count and then seven hours later still. Figure **6.3** illustrates the results. Before our eyes, the ratio of white cell counts in the two groups reversed over an 11-hour time interval.

This could be explained only if the two groups of mice had cycles in white cell count that peaked at different hours, as shown in Figure **6.4**. The rhythms for the two groups crossed each other around 10 a.m. At the time they crossed, the count was falling rapidly for the group of treated mice and rising rapidly for the group of untreated mice. If we made measurements before the crossover time, the treated mice would have a higher white cell count. If we measured after the crossover time, the untreated mice would have the higher count. If we measured near the crossover time, the groups would have about the same count. The three experiments had shown such different results just because of the few hours' difference in time of measurements.

But why were the rhythms for the two groups different? The cause of the difference turned out to be very simple. The un-

Figure 6.3 Comparison of the average cell counts in samples of the same groups of mice taken three different times; the ratio of the white cell counts reversed from week one to week three.

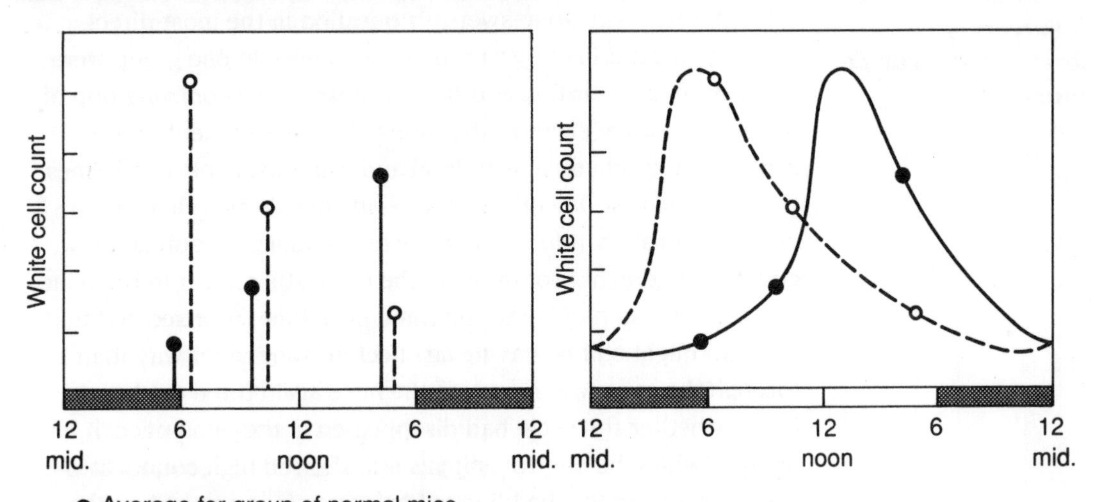

Figure 6.4 Speculated rhythms in white cell count for the two groups of mice. The different hours of peaking explained the puzzling reversals in cell counts observed in the earlier experiments.

treated animals had been free to feed at any time, while the treated animals were offered their special food only during a few hours of the day—hours when they would not usually have eaten. The feeding hours were convenient for the researchers but not for the mice, which prefer to eat in the dark. It turned out that the unusual feeding time of the treated group of mice had shifted the timing of their rhythms by about half a day. The answer to the second puzzle was that mice of the same species could have different rhythm timings, depending on *several* aspects of the conditions under which they were kept.

So we had learned that it was not enough to compare two groups at the same time of day. One had to be sure that body time was the same for both groups being compared, and that meant controlling the feeding schedule as well as the lighting schedule. Soon we were to learn that body time can differ for still other reasons.

A Puzzle of Light

We began to be more concerned about what caused the cycle in white cell counts. We had already learned that light had an effect, but the response to changes in lighting schedule took several days, so it was clear that we were not dealing simply with direct effects of light on white cell count.

To distinguish between the specific effects of blinding and the possible general effects of surgery, the control group of mice received mock surgery—anesthetic and a cut on the forehead.

An obvious question about the mice was whether the effect of light operated through their eyes or in some more subtle way through their skin. To answer this question in the most direct manner, an experiment was run in which mice in one group were blinded by an operation and then compared to a control group of mice with normal sight. Both groups of mice were kept on the same lighting schedule, with food and water available at all times.

For the first day or two, the blind animals as well as the control animals continued to show high counts at noon and low counts at midnight. Not only did the cycle still appear in the blind mice, but it also had about the same peak time. So it seemed that the light might still be having an effect in some other way than through the eyes. We measured the mice again three weeks later to see whether the cycle had disappeared in the blind mice. It had not. But while the control animals still showed high counts at noon, the cycles for the blind animals were found to be just the opposite—they showed high counts at midnight and low counts at noon! The timing difference between two groups had been found before, when they had different feeding times. But these groups were kept under identical conditions and at first showed the same peak times. Could it be that the peak times for the mice had shifted progressively over the weeks? And, if so, why?

Another experiment was needed to track more precisely what was happening to the blind mice. We already knew that nicking the tails of the mice to take blood samples lowered their white cell count, apparently because it frightened them. This unintended effect of measuring complicated the interpretation of results. We needed to measure something else that would tell us about the mouse rhythms—something that had a similar cycle to the white cell count rhythm but that could be measured without upsetting the mice so much.

An ideal variable would be one that could be measured quickly, precisely, and often, so we could repeat the measurements several times per day. Rectal temperature was chosen as an easily measured variable that might indicate the general physiological state of the animals.

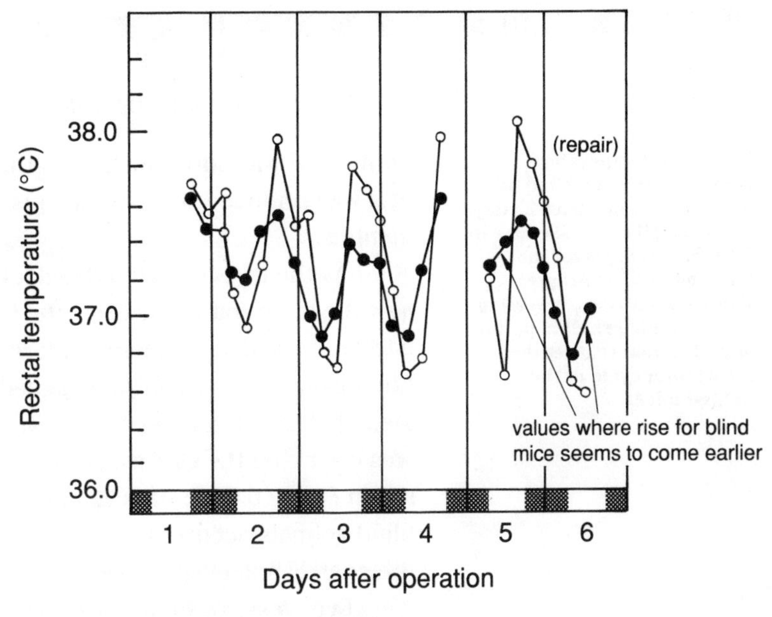

Figure 6.5 Average temperature measurements for groups of blind mice and control mice over six days. The last values suggest that the blind mice may have begun to show an earlier rise in temperature.

For the temperature experiment, mice were divided into a blind group and a control group and kept on the same lighting and feeding schedules. The rectal temperature of every mouse was measured with a tiny temperature-sensitive probe every four hours, day and night, without interruption (except for a short gap between days four and five). For the first few days the pattern of change in average temperature was nearly identical for the blind animals and the control animals. Then on the sixth day the probe broke. A very small difference in pattern between the groups on days five and six suggested that the temperatures of the blind mice might be starting to peak a little earlier (see Figure **6.5**). We decided to repair the instrument and continue measuring.

However, the director of the university department in which I worked was very skeptical and told me that the trend I thought I saw starting was only in my imagination. He did not want me to

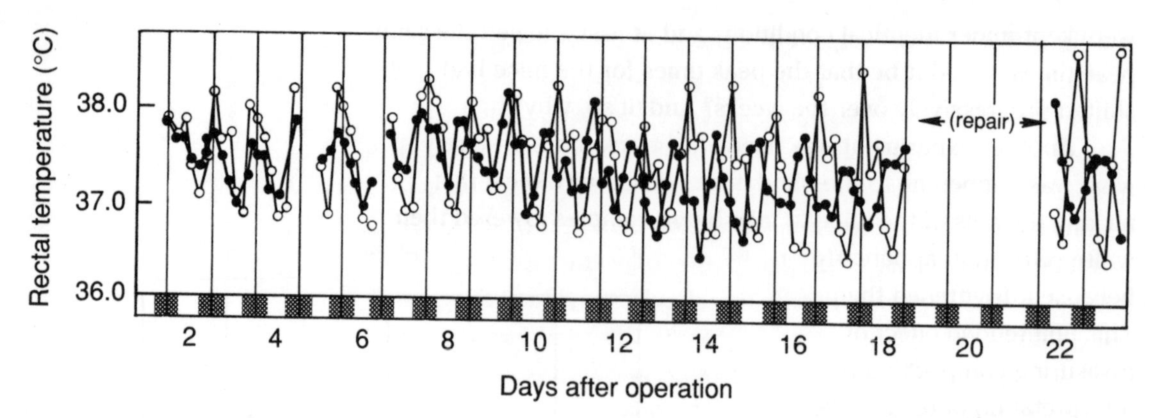

Figure 6.6 Temperature measurements for blind mice and control mice over 23 days (including the first six days in Fig.6.5). Averages for the blind mice show progressively earlier rises, revealing a free-running rhythm period of 23 1/2 hours rather than the 24-hour cycle of the sighted mice.

continue. I was sure we were on the trail of something important, and we continued the experiments in another laboratory, far off campus.

As the days of round-the-clock measurements progressed, we saw the temperature peak for the blind animals come earlier and earlier in the day. The temperature rhythm of the blind mice was not locked into the light/dark schedule—it appeared to be running steadily faster than one cycle every 24 hours. By the 22nd day it was clear that the cycles had become exactly opposite. As you can see in Figure **6.6**, on days 22 and 23 the peak of temperature in the blind animals occurred at the time of the lowest temperature in the control animals, and vice versa.

Even a series of measurements around the clock for several days had not at first shown what was happening. While the control mice were kept on a 24-hour cycle by the 24-hour lighting schedule, the blind mice (who couldn't perceive the light/dark change) ran on their own internal cycles of about 23 1/2 hours. We had learned that even if we made laboratory conditions as much the same as possible, we could not guarantee reliable results because the animals might be running on their own internal schedules. The answer to the third puzzle was that animals, if deprived of outside time clues, can run on their own internal cycles, which are not necessarily the same as their cycles in their usual environment.

Further Pursuit of Rhythms

To really pin down the behavior of the free-running cycles, we proceeded to make measurements on the mice every four hours, around the clock, with few interruptions, for the next two years. We continued the measurements, with longer interruptions (and other mice), throughout the 50s.

By now, as you might guess, our main interest had shifted

from the effects of hormones to the puzzling cycles themselves. What had seemed to be irregular results were actually important rhythms that had to be taken into account. We also learned that environmental schedules had to be kept similar to avoid misleading results. And we had seen that not even careful organization of schedules could assure meaningful results, because living things have their own internal schedules. In the following years we explored a wide variety of physiological cycles and how they ran. Today, thanks to the work of many scientists here and around the world, we know a great deal about biological rhythms and their importance in research and health.

Rhythms proved to be far more complex than we had imagined. They were found to have a variety of periods, including hours, days, weeks, months, and even years. Moreover, rhythms were found to affect one another in subtle ways, even when they had very different periods. Along the way, we learned that it was essential to analyze data statistically to detect and track rhythms, rather than rely on simple visual inspection of data. The adventure continues still with ever new puzzles in discovering rhythms, figuring out how they work, and deriving their implications for health and performance.

CHAPTER SEVEN

Activities in Chronobiology

This chapter includes several hands-on activities that allow you to explore some of the principles that you have learned in this unit on chronobiology. (Activity 8 is designed as a teacher demonstration, *not* as a student activity.) Some experiments in chronobiology can be relatively uncomplicated and may be performed using a minimum of materials. Some of the easiest and most interesting rhythm investigations can be done with your body as the laboratory. Perhaps you can use these activities as a guide in helping you design some experiments of your own.

ACTIVITY ONE

Running Hot and Cold

One of your body's most easily demonstrated rhythms is your inside temperature. Commonly believed to be a constant 98.6°F, your body temperature does not vary more than a degree or so, but it is seldom exactly 98.6°F. Instead you will find that your temperature has a distinct rhythm, with fairly predictable peaks and troughs.

Setting Up Your Experiment

All you need is a clock that indicates seconds and a thermometer that you can read easily. A digital thermometer is safer and easier to read than a mercury-in-glass one. Clean the thermometer using alcohol and cotton balls, then settle into a comfortable position in sight of your clock.

Equipment is not the only necessary preparation, however. As both the subject and the scientist, you must standardize *your* condition before making your measurements. You must not do anything that might alter the temperature of either your mouth or your body before measuring. Avoid eating, drinking, singing, or even talking for a while before measuring to keep the temperature of your mouth consistent with the rest of your body. You should also avoid strenuous physical activity which could affect your whole body's temperature.

Procedure

To measure your internal temperature, put the tip of the thermometer next to a blood vessel that is close to the surface. One of the best places is under your tongue, where there are two relatively exposed blood vessels. Place the thermometer under your tongue as close to the center as possible, close your mouth around it, and begin timing. When one minute has passed, remove the thermometer and take a reading. (If you use a mercury-in-glass thermometer, wait 5 minutes before taking a reading.) Record your temperature to the nearest tenth of a degree, i.e. "98.1°F."

When to measure

To get clear evidence for a rhythm, try to measure as often as possible over four to seven days. Once every two hours or so should be enough. A few measurements made in the middle of the night would also be helpful in revealing a more complete pattern. (If you decide to include nighttime measurements, be sure to wake up enough to measure accurately!)

Graphing results

When you have taken your measurements, plot the data on a graph with the hours of the day along the x-axis and the temperature (in tenths of a degree) along the y-axis.

Once your points are plotted, draw a line between the points on the graph to help you see the pattern more clearly. Without actual measurements being made, you can't be certain of what happened between the plotted points on your graph, so don't take the line too seriously. Also bear in mind that body system changes are much more gradual than the straight line connections of your graph might make it appear. To represent these gradual changes, you might prefer to use a smooth curve, without any sudden bends, rather than straight lines. This curve won't represent what actually happened either, but it will likely reveal the underlying pattern better than the straight lines.

For Further Study

Now that you have graphed your own temperature rhythm, you might like to compare it to someone else's. Plot their measurements on the same graph as your own—only use a different color pencil to differentiate more clearly between the two rhythms. Your temperature rhythm might peak at distinctly different times than theirs.

Because your body temperature rhythm usually corresponds to other rhythms related to measures of alertness (such as reaction time and coordination) people with different peak times in their temperature rhythms often have different patterns of alertness. A "night owl" and an "early bird" might have very different temperature patterns.

Another possible experiment would be to see the effect of strenuous exercise on your temperature. First, take your temperature as before, and record it to the nearest tenth of a degree. Exercise for five minutes, and take your temperature again. Then measure it again every five minutes until it returns to the first temperature. Plot your measurements on a graph with time on the x-axis in increments of five minutes and body temperature on the y-axis. How long does it take your body to return to your normal temperature (for that time of day, of course)? Does returning to normal take more or less time at different hours of the day?

ACTIVITY TWO

Heart Rate

Perhaps the most familiar rhythm in your body is that of your heart beat—the repeated contraction of the heart muscles pumping blood through your arteries and veins. The period of the cycle in which your heart contracts and relaxes is usually about a second or less. When you use your skeletal muscles vigorously, the heart pumps more rapidly, two or three times each second. Strong emotions, such as anger or fear, may also cause your heart to pump faster.

Even under normal circumstances, the rate at which your heart contracts varies from one moment to the next, depending on your activities and thoughts, as well as an underlying daily variation.

Setting Up Your Experiment

To record your heart rate, you need to be able to detect each beat. This is done most easily by using your finger tip to feel the swell of blood in a large artery as the heart beats. The best places to feel this "pulse" are on your wrist bones just below your thumb or alongside your windpipe in your neck, places where a large artery is close to the surface. Choose the spot where you can feel the pulse most clearly.

Once again you need to try to be in the same situation each time you measure because your heart rate is so sensitive to what you are doing and thinking. Begin by sitting quietly for a few minutes and thinking about something peaceful.

Procedure

When you can feel your pulse clearly and you are ready to begin, start the timer—or wait until the second hand on a clock or watch gets to 12. Count the pulse beats until a minute is over.

For Further Study

You know that physical activity and emotion affect your heart rate, so you might want to investigate how quickly it returns to normal after exercise. Take one measurement at rest; then run up and down a flight of stairs to get your heart rate up to about twice its resting value. Measure again. When you have rested for two minutes take another measurement. Record the difference between this last value and the first. Try this experiment at several different times of day. Does your recovery from a fast pulse show an underlying variation with time of day?

ACTIVITY THREE

How Time Flies

Your sense of how fast time is passing is related to your general alertness. When you are sluggish yourself, you imagine time to be passing very slowly. When you are alert, however, time seems to pass quickly.

You can perform many more experiments with biological rhythms in the human body. Instructions and equipment for other self-measurements (such as grip-strength and reaction time) are available in the *Body Scientist* kit, sold in nature/science stores nationwide by Bushnell. For further information write: Bushnell Co., 300 N. Lone Hill Avenue, San Dimas, CA 91773.

Setting Up Your Experiment

A simple way to check your personal sense of time is to estimate how long a minute is—without looking at a clock. Clear the area where you will take your measurements from distractions: The area should also be free from any time clues like ticking clocks!

Procedure

Set a stopwatch to zero. Then start it and turn it over so that you can't see the time. As you start it, also start estimating a minute. (To help you estimate, you might count to 60, trying to make each count equal one second—or imagine that you are watching the second hand of a clock go around once.) Just when you think a minute is up, stop the stopwatch. Record the actual time elapsed. Plot the measurement on a graph with the x-axis as the time of day and the number of seconds passed as the y-axis. Repeat the experiment every couple of hours over four to seven days.

An alternative to using a stopwatch is any clock that indicates seconds. Start estimating just as a new minute begins, look away while you estimate a minute passing, then quickly look back at the clock and record the number of seconds actually passed.

Remember, the important thing is not how close your estimate comes to a minute—but how much your estimate *varies* at different times of day—so be careful not to compensate for yourself in order to guess closer to the actual time. For example, if you find that your estimates tend to be too long at some time of day, don't try to "correct" by stopping a few counts early. Your estimates are likely to improve with practice anyway, so within a few days you may see a trend toward true 60-second minutes on your graph as well as a daily variation.

ACTIVITY FOUR

How Flies Time

This activity is designed to demonstrate that living things show about-daily rhythms in their susceptibility to environmental hazards. Specifically, the experiment shows that the survival time for groups of flies exposed to cold or poisonous vapor is distinctly different at different times.

Materials

100 fruit flies, as similar as possible, preferably the same age, sex, and type. Vestigial-winged flies should not be used. Assorted wildtype flies will do, however.
12 stoppered glass test tubes (approximately 20-ml test tubes)
Cotton
Fine steel wool
Ice
Alcohol or acetone
Clock with a second hand

Setting Up Your Experiment

For a few days before and during the experiment, the flies should be kept in a controlled environment with constant temperature (± 2°C), continuously available food, and a regular lighting schedule (for example, lights on 6:00 a.m., off at 9:00 p.m.). Several days of this should ensure that their rhythms are reasonably well synchronized. For each event you will prepare two identical test tubes of flies. One test tube will receive a treatment, the other will be a control, sub-jected only to the handling and tube conditions but not the same treatment.

Procedure

To examine the flies' reactions to cold, place eight flies in each of two test tubes and stopper them. Immerse one test tube in a clear-walled container of ice water or melting snow (Figure **7.1**). Record the exact clock time. As a comparison, keep the other test tube under normal room conditions, in the same position—so that the cold will be the only difference likely to affect the flies.

To examine the flies' reactions to toxic fumes, place eight flies in each of two test tubes. Immediately add a small plug of fine steel wool to act as a buffer between the flies and the source of the vapor. Place cotton plugs in each test tube. For one test tube,

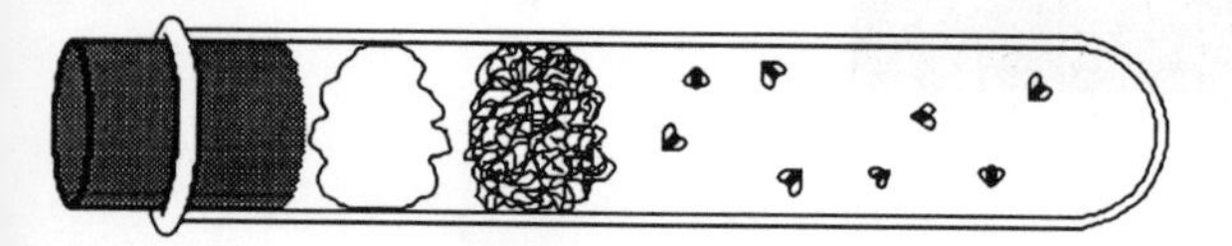

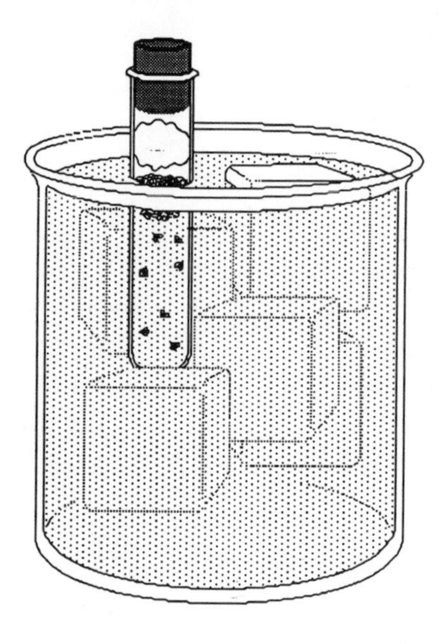

Figure 7.1 Apparatus for the study of cold's effect upon fruit flies at different hours of the day

place 10 drops of alcohol or acetone on the cotton plug. Quickly stopper the test tubes and record the exact clock time. As a comparison keep the other test tube nearby in the same position.

For either treatment, watch each test tube carefully. When the fourth fly drops in each, record the exact clock time. If possible, have two students work at each test tube—one to watch the flies and one to watch the clock. The comparison test tubes serve as "controls," to see whether anything besides cold or fumes might be causing the flies to drop. If any of the flies in the control test tubes fall, consider your fly-handling methods, the cleanliness of the test tubes, or any other factors that you can imagine might be affecting the flies. Then run the experiment again from the beginning with better conditions. (Probably nothing will happen to the control flies, but no experiment will be convincing unless it includes controls.)

When to measure

The best schedule calls for doing tests every four hours, around-the-clock, for at least two full days. As most school days cannot accomodate this schedule, try to space your measurements at regular intervals encompassing the widest possible span.

Start time	Clock time for four flies down	Survival time
8:00	08:01:52	1:52
12:00		
4:00		
8:00		
12:00		
4:00		

For practical purposes in the classroom, you can modify the schedule to three tests (at approximately 8:00 a.m., 11:30 a.m., and 3:00 p.m.) on each of two days. Graph results as in earlier activities.

ACTIVITY FIVE

Run, Rodent, Run

This exercise is a study of the activity cycle of a rodent, as measured by its daily wheel-running. Materials have been kept to a minimum, and you may use the apparatus to investigate a number of aspects of environmental influence and internal timing of circadian rhythms.

As always, when using animals in a laboratory procedure, maintaining good health and providing optimal care—based on an understanding of the life habits of the species used—is of primary importance. Supervisors and students should be familiar with literature on care and handling of living organisms. Practical training in these techniques is encouraged.

Materials

A hamster, white rat, or mouse, six weeks or older, available at most pet stores.

A cage with a running wheel. The wheel is preferably a part of the cage, but may be free-standing. (If you use a free-standing wheel, you'll need to attach it firmly to one side of the cage. The wheel itself must not have much horizontal play. You can reduce play by wrapping masking tape around the ends of the axle outside the supports.)

A 4-digit counter, either electrical or mechanical, preferably resettable. Electrical counters should operate on 6, 12, or 24 volts (rather than 120 volts), and thus will require a power supply.

Wire, for connections and contact with the wheel, if an electrical system is used.

A fluorescent light fixture, 15–40 watt.

A plug-in timer, for automatic switching of the light.

A large box or light-sealed room, to contain the cage and light fixture.

Setting Up Your Experiment

Figure 7.2 shows the basic setup: Use a counter to measure rotations of the wheel and, by reading the counter at intervals, draw activity graphs as a function of time. (If an automatic event-recorder is available, use it to provide round-the-clock data without the need for periodic readings.)

If you use a simple electric contact between battery and counter, solder or clip a wire to the cage (or directly on the stand of the running wheel if you use a free-standing wheel). The wire should not run through the inside of the cage, as the animal will tend to chew on it.

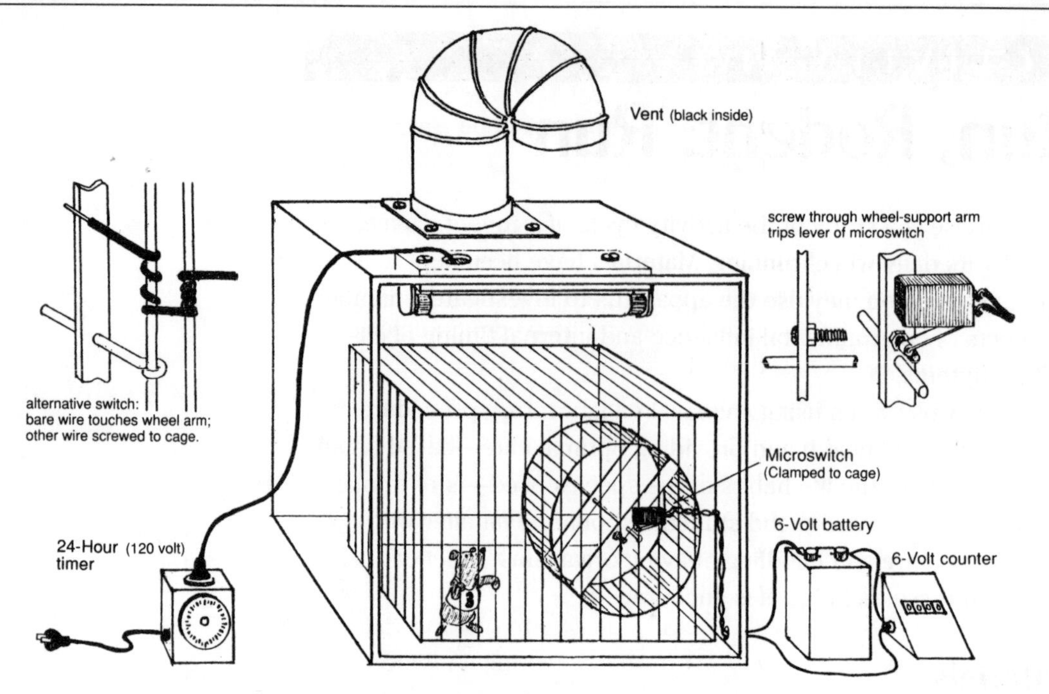

Figure 7.2 Apparatus for the study of rodent running

Use another wire as a contact with the running wheel. This wire sticks into the cage and brushes against the metal spokes on the side of the running wheel, completing the counter circuit twice with each revolution. You can secure this wire simply by weaving it through the bars.

Running wheels and cages are sometimes coated with protective lacquers that act as electric insulators. If you use one of these cages, sand off this covering where the contact strikes the wheel.

An alternate system is to connect the battery and counter through a microswitch. Secure the switch to the side of the cage so that something attached close to the hub of the wheel hits the switch lever with each revolution. (You can use the metal spokes on the side of the wheel, in which case the counter will register twice for each turn of the wheel.)

If you use a box to cover the cage, it should be large enough so that heat given off by the light does not raise the temperature of the box much. Rodents will not run if the temperature is greater than around 27°C (80°F). A small fluorescent fixture is best. The box may also be raised slightly off the ground and vented at the top; however, do not allow light to enter the box. Construct a light-proof vent by placing a curved tube with a black interior over a hole in the top of the box (see Figure **7.2**).

You can also use a small room or closet to contain the cage. If you use a room, it must be well ventilated but completely devoid of light from the outside. As always, be careful not to disturb the animal during the course of the experiment.

Connect an electric activity counter to a long pair of wires and place it outside the box or room so that it can be easily read without disturbing the animal.

Procedure

Make sure the animal has plenty of food and water available at all times, to minimize the disturbance caused by feeding. Replenish food and water only when the animal is awake—that is, when the fluorescent light in the box is off.

Feed and water the animal quietly and only when the light inside the box is off.

Place the animal in its cage in the box or room, with the light-timer set to turn the fluorescent light off at 9:00 a.m. and on at 5:00 p.m. Rodents will tend to run off and on during the entire dark period. If you can read the counter only during a six- to eight-hour school day, begin the readings about an hour before the animal's activity begins. This will make it easier for you to observe delay or advance in onset of activity under changed lighting conditions. Allow a week for the animal to become accustomed to its new environment and time schedule.

Recording data

At the start of the experiment, set the counter to zero at 8:00 a.m. Read the counter on a regular schedule—every half hour, if possible, but at least every hour. After each reading, reset the counter to zero. Enter the number on the data sheet along with the time it was taken and the date (see sample data sheet below in Figure 7.3). If several different people are recording counts, put your initials next to each recording, so that you will be able to

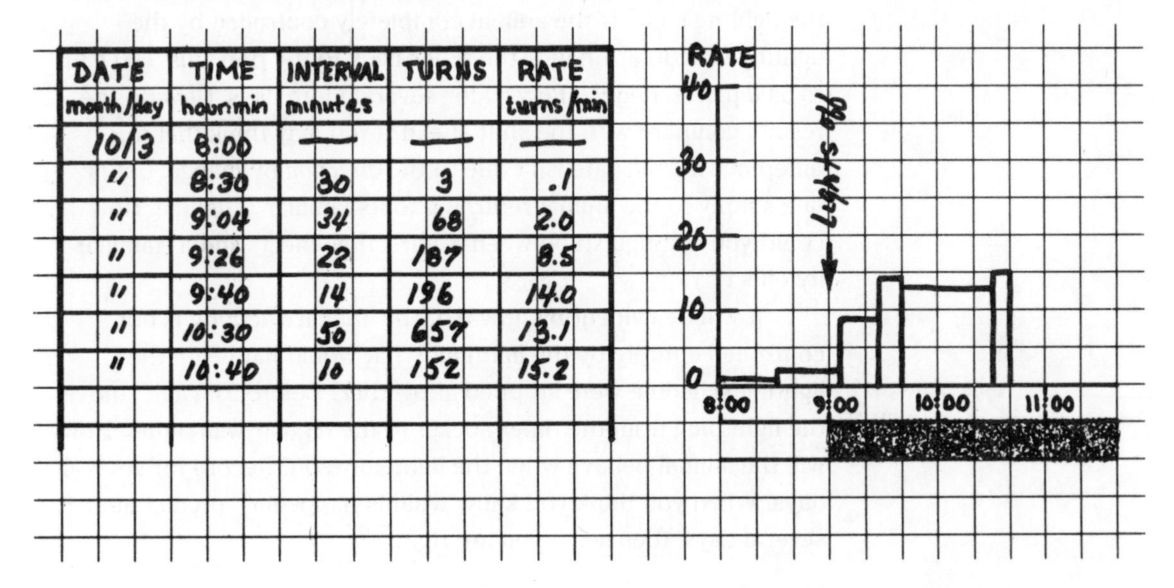

DATE month/day	TIME hour:min	INTERVAL minutes	TURNS	RATE turns/min
10/3	8:00	—	—	—
"	8:30	30	3	.1
"	9:04	34	68	2.0
"	9:26	22	187	8.5
"	9:40	14	196	14.0
"	10:30	50	657	13.1
"	10:40	10	152	15.2

Figure 7.3 Example record of the wheel-running activity of a hamster and a corresponding graph of average running rates

explain any peculiarities later. Leave the counter running overnight so you'll know how much running occurs during the animal's rest span.

Graphing results

Graph your data. Plot the time of each reading on the horizontal scale and plot the running rate (in turns per minute) on the vertical scale. Compute the running rate by dividing the number of turns by the number of minutes it took to make the turns (that is, the number of minutes since the previous reading). Draw each bar on your graph so that its right side is on the actual time the reading took place and its left side is adjacent to the preceding bar. Make a bar on the graph for each time interval (see the sample graph in Figure **7.3**).

Two days of recording should show that, on a schedule of 16 hours light and eight hours dark, the animal begins to run shortly after onset of darkness and continues, with some interruptions, for about eight hours. Activity will cease around the time of the onset of light.

For Further Study

The animal's activity appears to be controlled by the light conditions. You can investigate whether this is really true by changing the light schedule suddenly. One morning before 9:00 a.m., adjust the timer by four hours to turn the lights off at 1:00 p.m. and on at 9:00 p.m. Record for several days.

Compare the graphs for the shifted schedule to those for the previous schedule. What happened to the animal's activity during the lighting shift? Is the animal completely controlled by the lighting schedule? Change the timer back to the previous 9:00 a.m. to 5:00 p.m. schedule. Record for several more days. How do the results compare with the shift ahead? (You may think that a difference in shift rate isn't due to the *direction* of change, but is due simply to the animal returning to its familiar schedule. How could you distinguish between a "direction effect" and a "familiarity effect?")

It will be evident by now that the animal's activity is not controlled entirely by the lighting. What would happen if the lighting schedule were stopped altogether? Before 9:00 a.m., move the light plug from the timer socket to the regular wall outlet. How will the animal behave when the light stays on? Record for several days. When you think you know what is happening, predict ahead several days, then see if you are right.

How flexible is the internal timing system? Could it be forced into a shorter day (6 hours dark/12 hours light) or a longer day (10 hours dark/20 hours light)? (These schedules would preserve the 2:1 ratio of light to dark.) Design an experiment to investigate whether a rodent can adapt to a cycle distinctly shorter or longer than 24 hours.

This investigation of a rodent's daily activity pattern provides some clues to the animal's life outside of the classroom. Why is the animal active in the dark? If, in the natural environment, this animal lived in a hole, would it know when darkness fell? As the length of a day changes from season to season, would you expect the animal spend more or less time being active?

Detailed Procedure for a 4-Week Experiment

WEEK I

Monday
Feed, water, start 8 dark/16 light schedule of 9:00 a.m. off, 5:00 p.m. on (Establishes animal on reversed night and day schedule)

Tuesday to Sunday
Let animal adjust to new light schedule; start recording at 8:00 a.m.

WEEK II

Monday
Start recording at 8:00 a.m.

Tuesday
Continue recording

Wednesday
At 8:00 a.m., move timer setting four hours later to 1:00 p.m. off, 9:00 p.m. on; continue recording

Thursday
Continue recording

Friday
Continue recording (Animal shifts to new schedule over approximately three days)

WEEK III

Monday
Continue recording

Tuesday
Continue recording; after 1:00 p.m., move lights-off time back to 9:00 a.m. off, 5:00 p.m. on

Wednesday
Continue recording

Thursday
Continue recording

Friday
Continue recording (Animal shifts back, probably at different rate from prior shift)

WEEK IV

Monday
Continue recording

Tuesday
Continue recording; before 9:00 a.m., unplug timer and maintain continuous light in cage

Wednesday
Continue recording

Thursday
Continue recording (Animal drifts away from 24-hour schedule)

Friday
Continue recording; predict future activity pattern

This experiment could continue for many weeks.

ACTIVITY SIX

Tip Twining

Many if not most plant processes are rhythmic. Bean plants are especially convenient for showing several different rhythms. In this experiment, you will examine the twining movement of the bean plant's growing tip.

Materials

2 to 4 bean seeds
1 plant pot or paper cup
A high intensity light such as a reading light
A chalkboard or cardboard box
A small stake, or other form of support

Setting Up Your Experiment

To grow pole-bean plants (*Phaseolus vulgaris* "Kentucky Wonder"), first place soil in a pot (such as a paper cup), stopping a couple of centimeters short of the top to leave room for watering. Plant two to four bean seeds into the soil approximately 2 cm deep. When the seedlings emerge from the soil, remove all but the best one from the pot. Grow the plant in an area with suitable lighting and temperature until the pair of leaves above the cotyledons are almost fully expanded (typically 2 to 3 weeks after planting).

Procedure

Even a fairly short study of pole beans will reveal a movement cycle by which the plants twine around their supports. Eight to twelve hours before beginning measurements, place the plant in the room where the measurements will be made.

In addition to the general room lighting, place a small reading light to the side of the plant so as to throw a sharp shadow of the shoot tip on a vertical surface (such as a chalkboard or the side of a cardboard box) next to the plant. As in Figure **7.4**, mark the position of the tip shadow with a small zero. Make similar marks

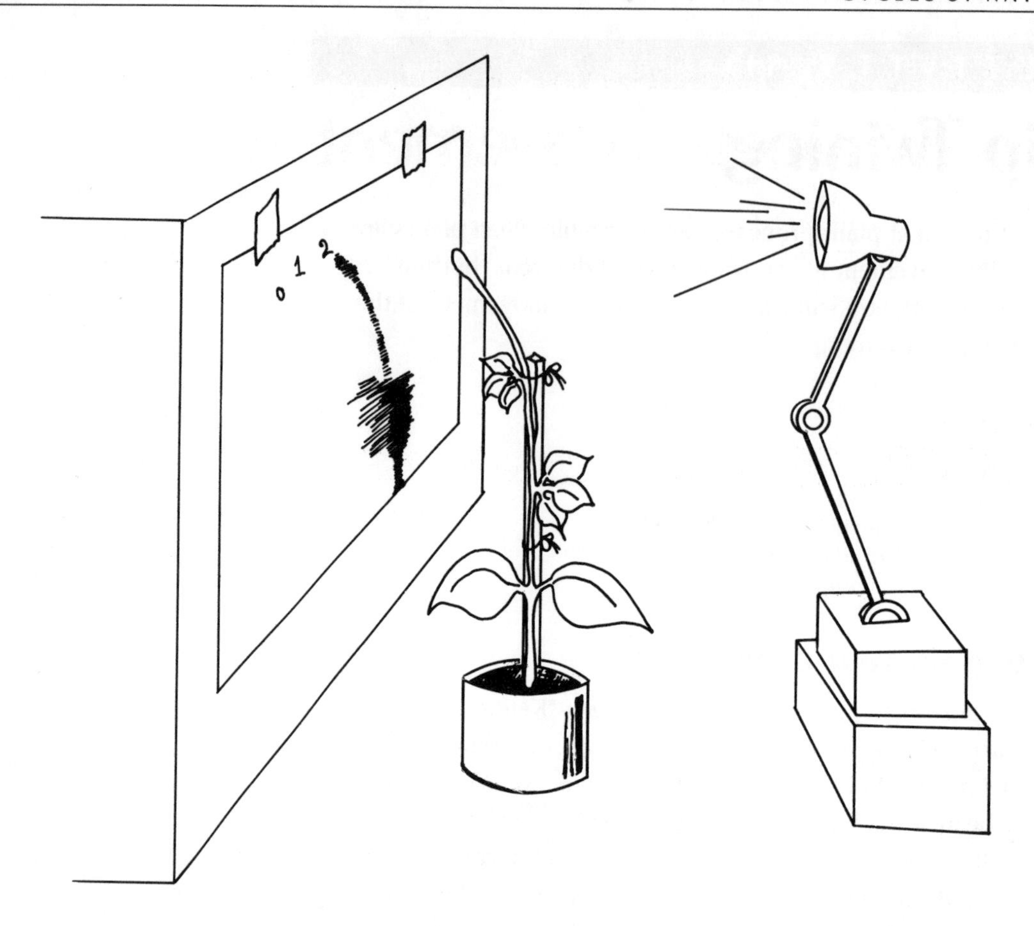

Figure 7.4 Method of measuring the growth of a bean plant

every 10 minutes for the next 3 hours or more, writing numbers 1, 2, 3, etc. for the successive positions of the shadow. Finally connect the numbers to show the pattern of the shadow's motion.

How does the shoot itself move? What is the period of the twining cycle? Is it driven by some cycle in the environment, or by some rhythm within the plant itself?

This activity has been adapted from "Chronobiology Projects and Laboratory Exercises," by Willard L. Koukkari, Jeffrey L. Tate, and Susan B. Warde in *Chronobiologia 14*(4) 405–442, October–December 1987.

ACTIVITY SEVEN

Plant "Sleep" Movement

A circadian cycle in up-and-down leaf movement is easy to observe in the largest leaves of the bean plant. This activity involves an experiment to see whether the movement is driven by the environmental light/dark cycle.

Materials

4 young bean plants
A fluorescent 15–20 watt light fixture
A plug-in timer, for automatic switching of the light
A stiff paper card with 10-degree gradations marked
A steel nut and string
A green "safe" light

Setting Up Your Experiment

Select 4 young bean plants that have fully expanded single leaves. Place the bean plants in a room where the 24-hour lighting schedule can be controlled by a lamp connected to a timer; a small 15-watt or 20-watt fluorescent lamp should be sufficient. (Incandescent light-bulbs may produce excessive heat and are therefore less desirable.) Set the timer to turn the lamp on for 16 hours and off for the remaining 8 hours. If the times for lights-on and lights-off are the same as those the seedlings were grown under, the leaf-movement monitoring may begin immediately. Otherwise, allow 3 days for the plants to become synchronized to the new lighting schedule before beginning measurements.

Procedure

When you are ready to begin measurements, move two of the plants to an area of continuous illumination in which the light level varies as little as possible. Plan to measure the leaf angles every 4 hours for a little over two complete 24-hour cycles (about 52 hours). This will probably require a team of people working together so you may wish to prepare a schedule detailing who will take the measurements at each time point.

Leaf angles can be measured conveniently with a stiff paper card that is marked every 10 degrees around its center and has a weight (such as a steel nut) suspended on a string through its center (to give a straight-down reference line). Approximate measurements to the nearest 10 degrees are satisfactory.

Ideally, leaf angles should be measured even during the schedule's dark span. This provides some difficulty; however, because exposing the plants to light even briefly during the dark span can shift the leaf-movement rhythm. For measuring leaf angles in the dark span, use a dim, green "safe" light, which does not appear to affect the rhythm.* If a safe light is not available, leaf angles should be measured just before and just after the dark span.

***Plants don't absorb much green light; they reflect it, which is why they look green.**

For all the plants, plot the measured leaf angles against time for two complete cycles. Does the cycle in leaf movement depend on light? What survival advantage would such movement give to plants?

This activity has been adapted from "Chronobiology Projects and Laboratory Exercises," by Willard L. Koukkari, Jeffrey L. Tate, and Susan B. Warde in *Chronobiologia 14*(4) 405–442 October–December 1987.

ACTIVITY EIGHT

A Colorful Revolution

This demonstration, mentioned briefly at the beginning of the book, is an exciting visual introduction to the study of cycles. The Briggs-Rauscher reaction is named after the two San Francisco high school teachers credited with its development, and it is generally regarded as one of the most impressive of the oscillating chemical reactions. Unlike the previous experiments in this section, the Briggs-Rauscher reaction is meant to be presented as a *classroom demonstration* by the teacher—it is not designed as a hands-on activity for students.

Materials

1 L 3-percent hydrogen peroxide, H_2O_2
29 g potassium iodate, KIO_3
1 L distilled water
8.6 ml 6.0 M sulfuric acid, H_2SO_4 (To prepare 1 L of stock solution, slowly and carefully pour 330 ml of concentrated [18 M] H_2SO_4 into 500 ml of distilled water. After the mixture has cooled, dilute it to 1 L with distilled water.)
10.4 g malonic acid, $CH_2(CO_2H)_2$
2.2 g manganese(II) sulfate monohydrate, $MnSO_4*H_20$
0.2 g soluble starch
20 g sodium thiosulfate, $Na_2S_20_3$ (See "Disposal" section for use.)
2 1-L beakers
hot plate
2 glass stirring rods
100-ml beaker
50-ml beaker
3-L beaker
magnetic stirrer, with 4-cm stirring bar

Setting Up Your Experiment

You will need to prepare three solutions before the start of your class. Solution 1 is simply your 1 L of 3-percent hydrogen peroxide.

To prepare solution 2, place 29 g of potassium iodate and approximately 400 ml of distilled water into a 1-L beaker. Add 8.6 ml of 6.0 M H_2SO_4 to this mixture. Warm the mixture on your hot plate and stir until the potassium iodate dissolves. Dilute the solution to 500 ml with distilled water.

For solution 3, dissolve 10.4 g of malonic acid and 2.2 g of manganese(II) sulfate monohydrate in approximately 400 ml of distilled water, using your other 1-L beaker. Heat 50 ml of distilled water to a boil in your 100-ml beaker. In the 50-ml beaker, mix 0.2 g

Hazards **The Briggs-Rauscher reaction uses and produces several potentially dangerous chemicals that must be handled with care.**

! Sulfuric acid is a strong acid and a powerful dehydrating agent that can cause burns. In the event of a spill, neutralize with an appropriate agent such as sodium bicarbonate ($NaHCO_3$), and then rinse clean.

! The reaction produces iodine in solution, in suspension, and as a vapor above the reaction mixture. The vapor or the solid is very irritating to the eyes, skin, and mucous membranes. In case of contact, flush the affected area with water for at least 15 minutes. If the eyes are affected seek immediate medical attention.

! Malonic acid is a strong irritant to skin, eyes, and mucous membranes.

of soluble starch with about 5 ml of distilled water and stir the mixture to form a slurry. Pour the slurry into the boiling water and continue heating and stirring the mixture until the starch has dissolved—this will take one or two minutes, and the solution may remain slightly turbid. Pour the starch solution into the solution of malonic acid and manganese(II) sulfate. Finally, dilute the mixture to 500 ml with distilled water.

Set the 3-L beaker on the magnetic stirrer and place the stirring bar in the beaker.

Procedure

You might want to use this demonstration as an introduction to a unit on biological rhythms. Therefore, this is a good time to raise some of the questions and topics that you will be discussing with your class over the next week or two. (Use the Introduction as a guide for your discussion.) Before or after the demonstration, be sure to explain the preparatory procedure that the class did not see.

Pour solution 1 and solution 2 into the beaker on the magnetic stirrer. Adjust the stirring rate to produce a large vortex in the mixture. Add solution 3 to the beaker. The mixture will turn amber almost immediately, and this color will gradually deepen. After 45–60 seconds, the mixture becomes a deep blue-black. The blue-black fades until the mixture is colorless, and then the cycle repeats several times. The period of the cycle is about 15 seconds at first, but it gradually lengthens. After several minutes the mixture will remain blue-black.

Disposal

The reaction produces large amounts of elemental iodine (I_2), which should be reduced to iodide ions before disposal. To do so, carefully add 10 g of sodium thiosulfate to the mixture and stir until it becomes colorless. **Caution—the reaction between iodine and thiosulfate is exothermic, and the mixture may become hot.** When the solution has cooled, flush it down the drain with water.

The description and procedure for the Briggs-Rauscher reaction is adapted from *Chemical Demonstrations: A Handbook for Teachers of Chemistry, Volume II* by Bassam Z. Shakhashiri (The University of Wisconsin Press, 1985), reprinted by permission of the publisher.

Bibliography

Aschoff, J., Ceresa, F., Halberg, F. (Eds.). (1974). *Chronobiological aspects of endocrinology.* Stuttgart-New York: F. K. Schattauer Verlag. (Report from Symposia Medica Hoechst 9)

Bennett, M. F. (1974). *Living clocks in the animal world.* Springfield, IL: Charles C. Thomas. (American Lecture Series #902)

Bünning, E. (1973). *The physiological clock: Circadian rhythms and biological chronometry.* (Rev. 3rd ed.). New York: Springer-Verlag. (The Heidelberg Science Library, Vol. 1)

Chance, P. (1987, October). The early bird makes the grade. *Psychology Today, 21*(10), 22.

Chapel, R. J. (1985, November). The ravages of time. *Nation's Business, 73*(11), 92.

Colquhoun, W. P. (Ed.). (1971). *Biological rhythms and human performance.* London: Academic Press Inc.

Conroy, R. T. W. L., and Mills, J. N. (1970). *Human circadian rhythms.* London: J. & A. Churchill.

Czseisler, C. A., Allan, J. S., Strogatz, S. H., Ronda, J. M., Sanchez, R., Rios, C. D., Freitag, W. O., Richardson, G. S., and Kronauer, R. E. (1986, August). Bright light resets the human circadian pacemaker independent of the timing of the sleep-wake cycle. *Science, 233,* 667–670.

Halberg, E. and Halberg, F. (1980, January–March). Chronobiology study design in everyday life, clinic, and laboratory. *Chronobiologia,* 7(1) 95–120.

Halberg, F. (1969). Chronobiology. *Annual Review of Physiology, 31,* 675–725.

Halberg, F. (1983). *Quo vadis* basic and clinical chronobiology: Promise for health maintenance. *American Journal of Anatomy, 168,* 543–594.

Halberg, F., Ahlgren, A., and Haus, E. (1984, July–September). Circadian systolic and diastolic hyperbaric indices of high school and college students. *Chronobiologia, 11*(3) 299–309.

Halberg, F., Drayer, J. I. M., Cornélissen, G. and Weber, M. A. (1984, July–September). *Chronobiologia, 11*(3) 275–298.

Hrushesky, W. J. M. (1985, April 5). Circadian timing of cancer chemotherapy. *Science, 228,* 73–75.

Koukkari, W. L., Tate, J. L., Warde, S. B. (October–December, 1987). Chronobiology projects and laboratory exercises. *Chronobiologia, 14*(4) 405–442.

Mills, J. N. (Ed.). (1973). *Biological aspects of circadian rhythms.* London: Plenum Publishing Company Ltd.

Moore-Ede, M. C., Sulzman, F. M., and Fuller, C. A. (1982). *The clocks that time us: Physiology of the circadian timing system.* Cambridge, MA: Harvard University Press.

Palmer, J. D. (1976). *An introduction to biological rhythms.* New York: Academic Press, Inc.

Peters, W. L., and Peters, J. G. (1988, May). The secret swarm. *Natural History, 97*(5),8–14.

Rentos, P. G. and Shepard, R. D., (Eds.). (1976). *Shift work and health: A symposium.* Washington, DC: U.S. Dept. of Health, Education, and Welfare.

Romanczyk, R. G., Gordon, W. C., Crimmins, D. B., Wenzel, Z. M., and Kistner, J. A. (1980, January–March). Childhood psychosis and 24-hour rhythms: A behavioral and psychological analysis. *Chronobiologia, 7*(1) 1–14.

Saunders, D. S. (1977). *Tertiary level biology: An introduction to biological rhythms.* New York: John Wiley and Sons, Inc., Halsted Press.

Scheving, L. E., Halberg, F., Pauly, J. E. (Eds.). (1974). *Chronobiology.* Tokyo: Igaku Shoin Ltd.

Seidel, W. F., Roth, T., Roehrs, T., Zorick, F., and Dement, W. C. (1984, June 15). Treatment of a 12-hour shift of sleep schedule with benzodiazepines. *Science, 224,* 1262–1264.

Shakhashiri, B. Z. (1985). *Chemical demonstrations: A handbook for teachers of chemistry, Vol. 2.* Madison, WI: University of Wisconsin Press.

Siwolop, S., Therrien, L., Oneal, M., and Ivey, M. (1986, December 8). Helping workers stay awake at the switch: Scientists are showing companies how to translate body rhythms into saner, more productive shifts. *Business Week* (2976), 108.

Van Bork, B. (1977) *Biological rhythms: Studies in Chronobiology.* 16mm, 22 min. Chicago: Encyclopaedia Britannica Education Corporation.

What time is your body? (1975) 16mm, 23 min. New York: BBC.